UBUNTU

The last of human freedom is the ability to choose one's attitude in given circumstances. This freedom no one can take from us. When we are no longer able to change a situation we are challenged to change ourselves.

~ Viktor Frankl, eminent Austrian Neurologist & Auschwitz concentration camp survivor.

UBUNTU

I AM BECAUSE WE ARE

PARABLES OF THE UNITED HUMAN SPIRIT

BASED ON REAL EVENTS

DR. SHILPA AROSKAR

ISBN 978-93-5458-830-3
Copyright © Shilpa Aroskar, 2021

First published in India 2021 by Leadstart Inkstate
A brand of One Point Six Technologies Pvt. Ltd.

123, Building J2, Shram Seva Premises,
Wadala Truck Terminal,
Mumbai 400022, Maharashtra, INDIA
Phone: +91 96999 33000
Email: info@leadstartcorp.com
www.leadstartcorp.com

DISCLAIMER: The views expressed in this book are those of the Author and do not pertain to be held by the Publisher.

Editor: Chandralekha Maitra
Cover design: Chinmayee Samant
Layouts: Kevis Tech

To

ANAND SHANDILYA

My dearest friend and *bookworm* buddy. This is for you. I hope, wherever you are, you will read and say with your trademark smile: *Isn't it fantastic?*

You are always fondly remembered in thoughts and action, like the words you once quoted so aptly: *The tragedy of life is not that it ends so soon but that we wait so long to begin it.* W.M. Lewis

CONTENTS

UBUNTU: I AM BECAUSE WE ARE

PROLOGUE

GOING TO WAR

I was in New Zealand when the first Covid case was identified in India. My family was a bundle of nerves about my return in case international travel was stopped and borders sealed. At the time I laughed at the fuss they made, telling them that I'd be happy to remain with the Maoris and pen novels in the lush green countryside. I was sure the Sars Covid-19 virus would be another Swine Flu or Bird Flu epidemic and would routinely vanish back to where it came from, in a few weeks,

How wrong I was.

Who knew that like a relentless inward tide, this spiky virus would flood every country, every continent, and expand its mutations to change the lives of millions, leaving mankind gasping for air?

Fortunately, I was able to get back home to India. A few days later, the first nationwide lockdown was announced, international travel banned, and borders sealed. Like a tsunami, Covid hit our nation. Lockdowns kept returning like a Netflix series, one after another. Together, we weathered the same storm, but not in the same boat. The rich, in the cocooned luxury of their abodes, tried exotic dishes and in-house gym routines, while thousands of migrant workers and daily wage earners walked hundreds of miles, ignoring the lockdown, to cross state borders and return home to their villages. Some drowned, others died of exhaustion, but the majority fought doggedly on for survival.

The fight was different for each of us – for air, for food, for medications, for beds in hospitals, for jobs, for income, for children's school fees – but there was one element common to all – the fight for 'freedom'. Freedom to return to BC (Before Covid), with no masks, no social distancing. Humans are social beings and very much emotionally interdependent for survival. People around the world were linked by a common fear, a common fight. The enemy made no distinctions.

Now, well into the second year of the global Covid pandemic, I sometimes wake and wonder if all this is real or just a nightmare. I had heard of the Spanish Flu epidemic, and read about the ravages of the World Wars, but I never imagined we would witness such an event in our own lifetimes, or be compelled to accept new realities.

Being a frontline doctor who works in Covid wards and ICUs, the last months have taught me more than any book of medicine or philosophy could have done. Neither the hospitals nor the healthcare workers were prepared for such an unprecedented crisis. However, in no time, every hospital and the entire medical community of doctors, nurses, ward boys, and community health workers, geared up like soldiers to save lives. Every day I returned home from the hospital with the fear and guilt that I might infect my own family. At work I witnessed every shade of human emotion, good, bad and ugly, behind masked faces, and the changing behaviour patterns of people from fright to flight to fight.

Amidst all this mayhem and chaos I tried desperately to look for positives, for small joys, for reasons to spread cheer at work, with friends and family. During the initial days of the first lockdown, there was little footfall in clinic practice. Patients were petrified to come out of their homes or visit hospitals. What could possibly be deferred, was.

For myself, when I am not inking prescriptions, I write books. Writing distils, crystallizes and clarifies my mind and brings solace to my heart. Following publication of my book YOLO, I received mail from readers enquiring about my next book. But there was never enough time, though I continued my humorous blog: Slice of my life.

Then, like a gift from the Universe, the concept of this book took shape amid the pandemic and lockdown. It was as if every person I encountered had an untold story to tell. This book is thus about weaving the lives and experiences of the various people I encountered in those dark days, from different spaces and professions – hair dresser, hotelier, maid, senior citizen, and many others – all bound by a common challenge – surviving the pandemic together. *Ubuntu* is about all coming together rather than as individuals. The pandemic has taught us one thing: *I am because we are*, the critical reality of our inter-dependency with each other and the entire ecosystem. It is what the ancient Hindu scriptures term *Vasudhaiva Kutumbakam*. It also redifined many things like luxury, freedom, love, friendship, festivals, e-schooling, and being able to find the small joys in challenging times.

We have all had to adapt to a new normal and learn to surf the Covid waves. What is happening to us is a delineating event in human history; what matters most is how we as humans, perceive and interpret it. Covid *will* end one day and much of what we experience now will pass into folk and medical lore, tales to tell our grandchildren. But every story has a meaning, perhaps a moral. It is time we understood the meaning of our story.

I hope that reading this book will help you look at life with a new perspective and unlock small joys, just as I did. May we all reboot our lives and let *ubuntu* and the human spirit triumph. Tomorrow is a new day.

THE WOUNDED HEALERS

*The planet does not need more successful people but
desperately needs more healers, restorers, peacemakers,
storytellers and lovers of all kinds. ~ Dalai Lama.*

The novel corona virus has transformed the
world and brought sweeping changes to
every sphere of life. The medical profession is no
exception. This pandemic will be remembered as
a revolution in the history of public healthcare.
On the one hand, it threw into harsh light the
inadequate healthcare systems of the world, even
in the most advanced nations, who were unable
to cope with either the virus or the host. On the
other hand, it demonstrated how the medical
fraternity and infrastructure could evolve
overnight, revamp and adapt itself to deal with
the ever-changing 'new normal' in OPDs, ICUs
and clinics. It demonstrated how people come
together in a crisis, and the resilience of the
human spirit.

Despite the many achievements of modern
medical science, there is no user's manual or guide
to tackle this pandemic. The medical fraternity

had no choice but to be *atmanirbhar* (self-reliant) and MIY (manage it yourself). It became a Tom and Jerry game between the virus and the host, with deadly consequences. Restaurants, retailers and small businesses were not the only ones to take a hit. Physicians too, witnessed a steep fall in private practice.

I often receive calls to attend emergency caesarean sections. Now, every pregnant woman who comes in for delivery, undergoes a Covid test as part of a routine protocol. In the BC era (Before Covid), such a summons to the OT would merely have entailed changing into OT scrubs. Attending a caesarean delivery rarely took longer than half an hour. Now it was like dressing to go into space. First one had to go to the donning area and put on PPE coveralls, masks, shield, goggles, gloves and stockings. Ideally, one required an assistant to help. The first time I put on the Covid costume, I struggled, but my Resident was quick to come to my assistance. And since we all looked the same, we played *Pehchaan Kaun* (Guess Who) in the OR. The other day, a surgeon asked me to count the gauzes thinking I was the OT Nurse. PPEs are certainly great levellers.

Another challenge is breathing. I wish there were better reasons to 'take my breath away' other than the N95 mask. Inside the PPE, one gets quickly drenched in sweat. Most annoying perhaps is that glasses get fogged up completely, impairing vision. But the doctor must go on — to operate, to intubate to treat patients, regardless. It is an exasperating and enervating challenge.

Clinics and OPD practice too, have undergone changes to cope with the unprecedented crisis. On one hand, frontline workers must worry about their safety, and on the other, take responsibility for the safety of every patient who visits the clinic. From installing sanitiser stands, to sealed triage areas, to spaced seating, to glass screens at the reception, clinics now look more like airports. Ironically, the fear of stepping out of the house has improved general health, due to enforced isolation, resulting in a dramatic fall in OPD footfall.

In the BC era, I rarely saw a patient without a prior appointment, nor had I the time to wait if a patient arrived late. But now, in the DC era (During Corona), it is quite the reverse. Doctors eagerly wait for patients to come, scheduled or not. In fact, I have so much time to kill in the OPD that if a patient tells me she is on her way, I not only wait patiently but also make some tea

UBUNTU: I AM BECAUSE WE ARE

to share when she arrives. Some patients prefer to wait in their cars rather than come into the clinic, so we call them in once their turn comes. Treating patients in the current time presents a whole new set of challenges for clinicians. Chief of which is MIY (Manage It Yourself).

Covid symptoms often overlap with those of seasonal flu and other viral fevers, so every patient with fever is unfortunately treated as a suspected Covid case. The entire clinic has also to be sanitised after seeing every patient running a fever.

It is difficult to recognise patients as everyone is now masked. A few enthusiastic ones even wear surgeon's gloves to visit the clinic, perhaps enjoying the chance to feel like surgeons without actually holding the scalpel. Being a Paediatrician, my tiny tot patients refuse to wear masks and a few hyperactive ones try to pull off my face-shield as well. It's all a game to them. The youngest ones have never known any other normal.

And now there's a new kid on the block — Telemedicine. In India, it has existed in a telephonic avatar for years, with patients phoning in with symptoms, some trying to get free consultations and visiting the clinic only if they did not get better. But the pandemic gave telemedicine a facelift. It is generally a paid-for service, with set guidelines of use. Rarely is the doctor live streaming. However, due to misconceptions, often generated by misguided advertisements, patients expect to see their doctor live on screen. *Grey's Anatomy, House, The Good Doctor* and other TV series have painted an unrealistic picture in patients' minds.

In my own case, I have sometimes worked from home, juggling household chores, and even done tele-consults from my kitchen. Some demanding patients even ask for a 'virtual exam', complete with stethoscope, no matter how unrealistic. After all, they say, they have paid for a full consult. Many of my overwhelmed patients wish to Facetime, and the whole family queues up to see what I look like in my PPE suit. Many tele-med users also think that since they have paid once, they can keep calling indefinitely to consult on every complaint thereafter.

What saddens me is that medical practice has lost its holistic approach amidst all this. Often, we make diagnosis without clinically seeing our patients. There is no healing hand of reassurance or compassion to offer. It feels robotic. The myriad emotions of patients remain unseen behind their masks – fear, anxiety, sadness, a smile or a tear.

Frontline doctors are no less than the Army; soldiers fighting for the country and its people. Many cannot go home for months, or hug their children or spouses. Dressed in PPEs they cannot eat or even use the washroom, often for 8-9 hours at a stretch. We do it gladly because that is the oath we took. It is what our conscience and humanity dictates.

And when healthcare workers do go home, life is no easier. The worry of infecting family members is all too real. We wash our clothes separately, and try to maintain social distancing, especially with the elderly. The emotional strain, the ethical dilemmas, the financial downturn, have all plagued the medical fraternity as much as the virus itself. But these are small sacrifices compared to the many healthcare workers and colleagues who have lost their lives to Covid. So when I see people acting irresponsibly by not wearing masks, not staying home when they can, crowding into liquor shops and eateries, or joining political rallies and religious rituals, I want to yell at them not to let the sacrifices made by healthcare workers to have been in vain.

Today's reality stems from the unpreparedness and under investment in the healthcare system. While enormous funds are spent on building temples and mosques, statues and plaques, little is invested in healthcare beyond kneejerk reactions to some crisis. Yet, right here, in the health sector, lies the potential to increase the national GDP. The Government must prioritise the creation of a robust medical infrastructure, and put in place sensitive action plans for when a pandemic or some calamity strikes. It is necessary to enhance the availability of medical education, going beyond quotas, so we have the army of doctors this country needs. The greatest lesson of the pandemic is that health is the greatest wealth.

If there is one place I have met God or experienced the miracles of the Universe, it is in my two decades of practice in hospitals and clinics. We must also heed the warning the Universe is sending us. We ask God to eliminate the virus, but it is also necessary to reboot our healthcare system and our thinking. Just as we are always prepared for war, we must be prepared for such pandemics, with better medical action plans, critical care units, isolation centres and special care for the elderly. The public and private health sectors must work in partnership.

The virus will eventually wane and vaccines will prevent its spread. But the long term effects and scars will take years to heal. The doctors and healthcare workers have borne the brunt of the crisis, emotionally, physically and mentally, but these wounded healers continue to serve, making the world a better place for all. They continue to mend the broken and heal the sick.

Perhaps, one day, there will be time to breathe.

MASTERING THE MASK

Man is least himself when he talks in
his own person. Give him a mask and he
will tell you the truth. ~ Oscar Wilde.

When the Hollywood super comedy *The Mask* was released in 1994, little did the writer, director or actor Jim Carey, envision that one day the whole world would be masked. Nor did actors Anthony Hopkins or Catherine Zeta Jones know that the *The Mask of Zorro* would re-emerge in 2020 as *The Mask of Sorrow.*

The only physical weapon we have against Covid, till such time as the entire population is vaccinated, is the humble mask. It is our best protective gear and has now become a way of life for all. No wonder fashion designers are busy creating designer masks. Designer Masaba even donated 1000 masks to the Mumbai Police frontline Covid warriors. Heeding the Prime Minister's call to be self-reliant, the middle class has used saris, *dupattas* and T-shirts, to create homemade masks, not only for themselves but for friends and the community. Many companies have created branded masks, and doubtless schools will do the same when they reopen. I wonder if these Covid era accessories will eventually be displayed in museums as historical memorabilia? For now, they are a reality of life.

Nowadays, it is a guessing game who is who behind the mask since the whole of mankind is masked. Perhaps we should wear ID cards. Women no longer need facials and lipstick, only eye make-up. The Ninja look is the latest trend. And those crowding liquor shops are perhaps at an advantage since no one can identify them. Every time I see my little patients walk in with masks over their faces, my heart bleeds. When will they once again run free, faces to the sun, wind in their hair, alive with the joys of childhood?

Apart from the feeling of suffocation, another nuisance of masks is the odour. *Airgasm* is a new word in the Covid dictionary. It signifies the relief of breathing in fresh air. How often we long to just rip off those masks and take a deep breath! The need of the hour is perhaps perfumed or anti-odour masks, perhaps with floral scents for women and musk for men!

Another nuisance is that masks get wet — from sweat, mucus and saliva. Dry-fit masks need to be manufactured, as well as anti-fogging ones. This would be a great boon to spectacle wearers. Those gastronomes addicted to constant consumption of food and drink, snacks and wafers, could have masks with flaps. Coffee addicts would be delighted with coffee-holes. Pet lovers would no doubt be glad to buy pet masks.

We Indians are innovative by nature. The *Kama Sutra* was after all written here. In similar vein, we have now invented the *Corona Sutra*, mask-wearing positions. The most popular is the necklace position, another being goggles-on-the-head. The lips-only, or the one-ear-mask, which dangles beside the face like an ear decoration, or those tucked into shirt pockets or sari blouses, all have their adherents.

For the best protection, however, one should wear a mask that is comfortable and well-fitting, like a good undergarment. It should be dry and clean, and changed frequently. And never borrowed or lent. Avoid adjusting it in public and do not drop it below the nose-bridge. Whether mask protocol will last for months or years no one knows. Who knew that just being able to breathe fresh air would become a luxury or a freedom to crave? But masks are our best protection and must not be ignored.

UBUNTU: I AM BECAUSE WE ARE

Having worn masks for almost two years now, people are experiencing a new kind of blindness – the inability to recognize people. The technical term is *prosopagnosia*, a neurological disorder that impairs the ability to recognize faces. In the pandemic, many of us have developed this problem. The other day, at a grocery store, I thought I saw a friend and loaded him with my groceries while I paid for them. The poor fellow, a perfect stranger, never said a word!

We all hope that when this ends we will unmask and re-emerge as human beings, revealing our true selves. Till then, flaunt your mask chic and smile with your eyes. Love your laughter lines.

THE ART OF BEING BORED
IN LOCKDOWN

Doing nothing often leads to the very best of something. ~ Winnie the Pooh

While the intelligensia are busy analysing Covid 19, and conducting esoteric debates over what the Government should do, others like me are busy analysing the Boredom Plague brought on by the lockdown. This particular ailment is more pandemic and virulent than Covid 19. It is necessary, as with all conditions, to get to the root of the problem. In this case – boredom.

What if I were to tell you that to do nothing is not laziness and apathy but the opposite? Boredom can be good, make you more productive. Why not cease to be part of the bragging competition to post fancy pics and hashtags on social media – new skills# lockdown productivity# unleashing talents etc. The wonderful truth is that even if you do absolutely nothing, you will be just fine. In fact, *being bored* is in itself a talent.

The word 'boredom' entered the English language in 1766. Tolstoy defined it as 'desire for desires'. Here are the types of boredom (yes, boredom has classifications too).

1. Indifferent: This type is seen in people who are calm and withdrawn from the world, relaxed or cheerful-fatigue types.
2. Calibrating: Characterised by wandering thoughts, these people do not know what to do with themselves. Common with those performing repetitive tasks.
3. Searching: Found in restless people who are actively searching for hobbies, leisure activities etc.
4. Reactant-aggressive: It motivates people to leave a boring situation and look for highly valued options.
5. Apathetic: This is a deeper negative state of mind linked to depression and helplessness.

Science says boredom is good for the brain. It improves creativity. Boredom serves as the lacklustre yin to our yan of excitement. It improves productivity. It also triggers within us the search for self-meaning, prompting us to charitable behaviour, and engagemnet with pursuits essential to sustained happiness. That being so, here are some tongue-in-cheek tips on: *How to be bored the right way.*

1. Rate your boredom on a scale of 1 to 10.
2. Tell yourself it is okay to be bored. Self affirmations help.
3. Pick boring activities that require little or no concentration, like strolling from kitchen to hall, rolling on the floor, or crawling, telling yourself they are quarantine workouts.
4. Binge watch movies or troll posts for hours.
5. Talk to pessimists who see negativity in all situations.
6. Write down 10 things you *don't* want to do.
7. Foster the practice of NDTA (no desire to do anything)
8. Lower the bar and set your goals low; get up late so half the morning is gone. Spend extra time on daily routines like brushing teeth or bathing. Sing in the bath. Spend extra toilet time, doodling perhaps if there are no newspapers.

9.	Encourage meditation – not you but your spouse, so s/he is calmer and you get that much more time to daydream.

10.	At family meal times, discuss boring topics like who discovered Mohenjo Daro, or the principles of flight, what Pandas eat, or which animal hibernates when. In my house we discuss cranial nerve pathways and brain anatomy.

11.	Call 'bored room meetings' on zoom.

12.	Don't solve Sudoku or play chess (that's challenging to the brain), instead, count the number of tiles, hinges, doors, windows in your house.

13.	Indulge in long cooking fests, the more inedible the dishes the better. Spend hours baking hard, not spongy, cakes.

14.	Read encyclopaedias, biographies, the atlas. Google a list of 10 most boring books and hit them.

15.	Read all the enlightening Covid Whats App messages from your learned friends and family, make notes, and then make them go viral. This is a jading task that will truly leave you pulling your hair.

16.	Mask making. Use *dupattas*, bedsheets or your wedding sari (if by now you have decided matrimony is an over-hyped state). Make it a fashion statement.

17. Once you are done with your boredom To Do list, exchange it with your peers to improve BQ (Boredom Quotient).

18. Set aside 'boredom time' in your calendar. A life full of thrills and frills can be exhausting, with you constantly on the lookout for stronger stimuli, akin to craving opiates. Take time out.

Some amount of boredom is essential for a happy life, else you will have a dull palate for small pleasures. Having written this chapter, I too, am feeling better. I will now close my laptop and only open it again when boredom woos me again.

Stay Bored. Do Nothing. You shall emerge bright and shining.

CON-QUESTING CO-RONA

Every sunset is an opportunity to reset.
~ Richie Norton.

The world metamorphosed in one short month, from 14 February, *Valentines Day*, to 14 March, *Quarantine Day*. 'Made in China' will never again signify an efficient duplication of stereo-typical products, for Covid 19 was the most powerful and original super-power ever to invest our planet. Should we raise a toast to human ingenuity or human survival? Ironically, is drinking alcohol now safer than eating? The truth is more alcohol has been consumed as hand-sanitisers in the last two years than has been drunk.

In these Corona times we all received umpteen guidelines for prevention and safety, thanks to social media universities. With malls, schools, colleges, and offices shut, everyone was left with no choice but to be at home. After all, we had to help flatten the Covid curve, even at the cost of our expanding waistlines.

However, the ones who managed to have a peaceful stay at home wisely followed certain guidelines:

1. Closing the social distance at home. Since keeping distance was normal, closing it was a challenge, especially for spouses committed to better or worse.

2. Many men attempted to regain their Dad titles by spending more time at home instead of making only guest appearances.

3. Rediscovery of spouses was on the cards but trying to understand them was as vain a task as it had ever been. A fallout of such bonding was the spilling of beans about girl/boyfriends past. Game over.

4. Husbands entered new territory – the kitchen. Wives gave TOTs (training) in cooking, home management, and most crucial, handling children. In my house, when my Ophthalmologist husband asked for an after dinner scotch, I handed him a scotch brite scrubber.

5. Children began cleaning their cupboards and making beds. They learned they had to earn their pizza and pasta by helping with household chores.

UBUNTU: I AM BECAUSE WE ARE

6. If the family did not comply with requests to help in the house, the House PM called a meeting at 8pm, in the living room, and declared two things:
    i.   Kitchen lockdown
    ii. Internet lockdown

    The family scrambled to help.

7. Family meal times were restarted, with fruitful dinner table conversations.

8. Those who spent time with their families made the startling discovery there was much to know and like!

9. The best immunity booster in lockdown was avoiding arguments with the spouse, as these led to stress, reducing immunity and making one more susceptible to infection.

10. The best way to avoid touching the face was to keep the hands occupied — cooking, cleaning, typing, holding two wine glasses.

11. One had to remember that lockdown would not last forever while buying groceries and essentials. So no hoarding. I saw women buying boxfuls of sanitary products, others hoarding batteries and

provisions that inevitably had a shelf life and would have to be thrown away in due course. What did one do with 25 batteries anyway? And thank god Indians prefer water to toilet paper.

12. While the whole world analysed the actions of various countries, there were those who preferred to analyse what our own Government was doing, extending unwanted advice on how to stop the Corona spread. The debate raged.

13. Women got busy on Facebook with the sari challenge, posting saree-draped images that no doubt spread cheer and envy.

14. Zoom became a saviour in these social distancing times. Sitting at home, we shared *Mann Ki Baat* with friends or escalated our learning curve with digital learning.

15. We took pride in teaching *namaste* to the world. In return, the West taught us WFH (work from home) and WWH (working without house-help).

There was no social mingling but was it really so bad? Some of us, having spent years acquiring

UBUNTU: I AM BECAUSE WE ARE

our homes, we finally had the opportunity to be at home. Ironically, how restless we became to get out of those same homes! The lockdown taught us what it felt like to be caged in like animals. It also taught us what freedom is.

Everything happens for a reason and we should always look at the glass as half full. The world has come closer. While international borders were closed, differences were forgotten in a common cause. The pollution levels went down. One woke to the sound of birds chirping and not cars honking. At last we had time to stand and stare at the sky, appreciate sunsets, trees and flowers.

We love to seek pleasure in restaurants, malls, theatres, gyms and pubs, but it's all so transient. The lockdown forced us to look within for deeper joys, to revel in the small things — that we still had milk for tea, food on our plates, chocolates and cookies to indulge in, movies and television to entertain us, the ability to exercise in our own homes to stay fit. It is a sad commentary on the human race that we needed a pandemic to realise all this and to be the catalyst for change.

It might take a long time to climb back up the mountain. The economy on the individual as well as national levels will take months, if not years to recover. But humans are resilient survivors. And there is always light at the end of the tunnel.

In the meantime, we should practice the 4Rs:

1. **Resolve:** To flatten the curve and co-operate to fight and eradicate Covid 19.
2. **Resilience:** While every individual has/is fighting his/her own battle, some on ventilators, some as suddenly single parents, some as jobless, we all need to maintain our stoic resilience. It's a test of human fragility versus endurance.
3. **Return:** Eventually this will end and we will return to old ways and routines like shaking hands, hugging friends, eating in restaurants and shopping in bazaars, malls and shops.
4. **Reform:** While there have been financial losses in every sector, some more than others, we have a common commitment to restore, rebuild and reform our homes, our lives, the nation, and the world.

Yes, this too shall pass. One day we will get up, don our office clothes and go to work. This time will fade, as nightmares do. In future years

UBUNTU: I AM BECAUSE WE ARE

we will undoubtedly narrate Covid 19 stories and quarantine folklore to new generations, about sanitisers, wearing masks, cleaning and scrubbing; of fears exposed, dreams crushed; of the hope that kept us going; of small acts of kindness, compassion and comfort that filled our hearts.

So laugh out loud as often as you can, fear less, help each other, and *live* more. Science and prayer, those unlikely bedfellows, have come together to reboot our lives. *Ubuntu* – together we overcome.

THE HOARDING BUG

Minimalism is a journey from the compulsion to consciousness, consumerism to common sense.
~ Amit Kaalantri.

People all over the world dealt with the pandemic and its collateral damage in various ways. One was the epidemic of panic buying, perhaps prompted by fear, uncertainty and anxiety. I was in a small departmental store buying essentials (of course, the definition of essentials is also as varied as people themselves), when a man piled dozens of batteries on the counter. I could not help but wonder what he could possibly want with so many? Covid 19 has certainly changed the way we think and behave, like hoarding provisions without thought for the next person.

The only difference in this hoarding mentality is that essentials vary somewhat across the world. While people in the UK hoarded toilet paper, in the USA it was frozen food and cooking dough. The French stocked baking and beauty products, and the Italians pasta. For Indians however, water is a cleanser and food in the stomach more important than food for thought. Before Covid, women of the Gujarati community routinely stocked quintals of grain for the entire year. During Covid, everyone began buying tons of grains and pulses, despite the PM's assurance that India had plenty of both. Instant noodles packets disappeared instantly from the shelves. People craved it like a drug. I saw people buying cartfuls of chocolate, as if Covid would lead to a chocolate famine and it was their last food for survival. Beard trimmers and beauty products were next to go as lockdown went into its second month. Grey became the new normal and trendy hair colour.

All this was merely a reflection of the uncertainty people felt all around them. Who knew what the morrow would bring?

However, Mumbaikars wisely invested in themselves as well. Condom sales were highest there, followed by the I-pill and home pregnancy kits. A baby boom was clearly on the cards. Good news for Paediatricians.

The Russians responded by stockpiling cash, withdrawing more and more money from ATMs, fearing they would soon be unable to access their money. They also kept their spirits high with spiking Vodka sales.

Essential goods did indeed stay in the market, though prices soared. But there were still those who chose to hoard whatever they could, including masks and hand-sanitisers. Surgical triple-layer masks and N95 masks went out of stock and became unavailable to doctors and healthcare workers. And if this was not enough, a few months later, people began hoarding drugs and oxygen cylinders used in Covid treatment. Even expensive drugs like Remdesivir disappeared. In one of life's great ironies, while patients desperately needed these lifesaving drugs and oxygen, there was an acute manmade shortage. Frantic relatives of dying patients ran helter-skelter, paying out entire life savings to buy these in a thriving black market.

Humanity was forgotten.

A businessman friend called me to ask whether he should buy Remdesivir and Tocilizumab for his family of four. He had kept aside 10 lakhs cash for this purpose, he said.

It was a stunning revelation about what was happening to us.

Thousands died; many unable to access or afford drugs and oxygen while the affluent and virus-free, stockpiled medications to reassure themselves. Self-preservation is perhaps the strongest trait of human nature, and compassion and empathy the first casualties in times of struggle. Hedonistic self-interest has ravaged the very soul of humanity like a metastasizing cancer. There can never be enough for the beast within.

Einstein rightly said that three great forces rule the world: Fear, greed, and stupidity.

Unfortunately the human race displayed all three.

The pandemic should have been a catalyst for change in our thinking, our behaviour towards each other, towards the ecosystem and the planet we share. Alas, our stupidity blinds us, and for all our human gifts of intellect, we have still to conquer a virus. The survival of the fittest has become survival of the richest. And the poor, as always, are the first victims of disease and death.

What Covid has taught us is the need to debug our minds and hearts and find the good within. Only then can we truly emerge victorious in any war.

It's time to learn that Less is the new More.

UNLOCKING SMALL JOYS IN LOCKDOWN

It's not the circumstances that create Joy. It is YOU.
Find the extraordinary in the ordinary. ~ Author

Awake since 5am, I could hardly wait for the 8am doorbell to ring. Should I make *elaichi chai* (spiced tea) and keep breakfast ready? What if the arrival was delayed? I finally leapt out of bed and brewed a cuppa *chai* for myself. I had dreamt of this moment in time (more than Whitney Houston perhaps), for two months of lockdown, every single day. I placed sanitiser, hot water, soap, apron, and mask ready. Finally, the doorbell rang and there she was, clad in a green sari, hair in a neat bun. Amita, my domestic help was back in my life. Tears of joy welled in my eyes as I resisted an intense desire to hug her. Social distancing, I reminded myself.

Small luxuries bring loads of happiness.

After 78 days, I finally sat in peace to sip my mocha coffee and enjoy my breakfast, reading the e-newspaper. During lockdown, mornings had been frenzied. As soon as I opened my eyes,

I would bolt out of bed, rush to the kitchen and start boiling milk on one ring, tea on another, and place the cooker on the third. Why had I never got a six-ring stove, I asked my better half? 'If you cook any faster we could probably open a restaurant or feed the whole housing society,' he replied.

*   *   *

'Mom I need some new clothes!' my daughter complained. I sighed. I had tried to avoid all but necessary shopping in these reduced times. Indian moms are a dab hand at making do. After all, only essentials work in Covid times, be it essential services or essential items. 'But Mom, these are basic needs,' she assured me. 'T-shirts, pullovers, nail polish.'

'But you're at home sweetie, so where are you going to wear all this?' I rebut. 'Mom, you just bought expensive hair dye to colour your hair. Besides you were the only one going to the hospital daily, even during lockdown, so I know you earned.'

Like a smartphone, she belongs to Smart Gen X. Too late to regret my maternal habit of feeding my children almonds daily for DHA (sharpens the brain). I give in and we do some online shopping.

❧❧❧

It's evening and I am brewing tea and ideas for my next book. My med-school-going son comes into the kitchen. Now this is a rare moment, I think. For two months he has religiously complied with the lockdown and remained in his room under cover of studying and doing online classes. Like a ferret, he emerges from his hole only when hungry. So, before he can ask me to fry hot *pakodas*, I hand him a packet of Marie biscuits. He surpasses his father when it comes to being a man of few words. In fact, he is a man of no words at home. He wears an expression only I can decode. I give in and tell him to order his favourite Belgian Waffles from Swiggy. He smirks and pulls my cheeks – a gesture of affection rarely displayed by man-sons.

The waffle box arrives and three of us gather round it as if Amazon had sent us a free iphone 12. The better half, however, is a contented soul and remains glued to his zoom webinar. 'You both got to give me half your waffles,' I tell the kids. 'It's your opportunity to pay back for supplying you both with food for nine months in the womb, and another year thereafter.'

Both offspring gawk at me as if to say, 'Mom you're the limit'. But I do get a bite from each. The hot, crispy, thick chocolate dripping creamy waffle fills up my senses like John Denver and I realise there are so many little pleasures we have missed in lockdown. A waffle never tasted so heavenly. Sheer joy.

🌢🌢🌢

I survey the pile of clothes on the bed and sigh. *To do or not to do* is the dilemma. Covid era dilemmas are different – whether to iron or not, to cook *khichdi* or Chinese, to scrub the bathrooms today or tomorrow. I decide to call the building watchman and enquire if he has seen a *istriwala* (ironing man). I yearn to be liberated from ironing duties. The watchman's networking and broadcasting abilities are enough to give Marc Zuckerberg a complex. In a jiffy he has an *istriwala* at the gate. I happily bundle up the pile of unironed clothes and give it to him at the gate. When he returns them, hot and crispy ironed, I feel as happy as if I had been given a designer sari.

🌢🌢🌢

On my way to the clinic I see a few shops open. The mannequins, draped in colourful clothes, seemed to wink at me and say, 'Hey, we're back in business!' I smile back and say, 'Glad to see you back.' My better half thinks I have finally lost my mind. From soliloquies to talking to mannequins. What next?

The rains are back. I lose my umbrella on the first day of the monsoon. My better half, who seldom misses out on such opportunities, points out unctuously that he has been using the same umbrella for six years, while I buy two every year and misplace them both. 'It's been 23 years with you and alas I haven't misplaced you yet,' I murmur as I search the whole house and finally find a broken umbrella. I am determined to get it fixed. After driving through every local lane and bylane in search of a cobbler, I am about to give up and buy a new umbrella afterall, despite the jibes, when I see a cobbler. I might have just spied George Clooney, so thrilled was I. Smiling as if he was a long lost friend, I ask if he can fix my umbrella. Silently, he nods. And lo it is done.

The lockdown deprived us of many things we took for granted, but it also taught us so many worthwhile lessons as well. It reminded us of the joy inherent in the small things of life. The happiness of indulging in little luxuries, like finding an *istriwala* or cobbler, getting back our domestic help, biting into a chocolate and cream laden waffle... The list is endless. Forget going on vacation to Tuscany, I would happily hop onto a bus for Mumbai *darshan* with my kids, or go for a drive over the Worli Sealink or along Palm Beach Road to my favourite Baileys outlet for a cup of hot chocolate.

Does this mean we have lowered the bar of what we expect from life? Or have we in fact raised our standards of contentment with small things? Perhaps both. The virus has been a definite catalyst for change in our lives. It is time to stop wallowing in self-pity and burst our fears like an overstuffed balloon. Let us instead look forward to unlocking our world with childlike curiosity and excitement, and embrace the new normal with cheer.

CORONOSUTRA:
LOVE IN TIMES OF COVID 19

The best thing to hold onto in life is each other.
~ Audrey Hepburn

Do not raise an eyebrow for this has nothing to do with the other *sutra*. While the world is seized by the Corona virus, with borders sealed and social distancing the norm, love and relationships have taken a hit too. Even Italy, the hub of eternal love, and Paris, the City of Lights, have lost their unique charm. Bouquets of roses have been replaced by bouquets of colourful masks, and perfume by sanitiser, as gifts of love. We are also getting used to a new lexicon: Covidial babies, Covidivorces, Cohooked, and Quaranteens.

Singletons vs. Married Coupledoms

For singles ready to mingle, the pandemic erased real romance from their lives. So they chose to spice up romance digitally and swipe right. Online dating spiked. Married couples, cooped up together at home, have probably had a lifetime of

UBUNTU: I AM BECAUSE WE ARE

toxic togetherness by now. Often finding it tough to live and let live, they constantly amplified every foible and quirk, be it hogging the TV remote or doing the household chores. And there were no getaways with friends, no escape. No doubt, social distancing of more than six feet was happily complied with for peace at home.

While Epidemiologists were still arguing about airborne or surface spread, we Indians had already discovered our own modes of transmission - touch, food-borne, sex, and even phone-borne. A patient frantically called me to say she wanted a Covid test done because she had spoken for 30 minutes with her cousin on the phone and the cousin had turned out to be Covid positive.

Paradoxically, there has been a baby boom. Though many Covid statistics have proved inaccurate, so much so that I no longer trust the numbers, being a chronic optimistic I do expect the baby boom to continue. I see people wearing masks around their necks or dangling from one ear, and I wonder if this is also the casual way they don contraceptives. No wonder India tops the list in contraceptive failure. Yes, definitely the baby boom is set to continue. For us Paediatricians, there will be bundles of joy to look forward to.

Love is Quarantined

In my house, we each quarantine in our own rooms. However, to foster family bonding, we meet in the hall twice a day. Mostly it's just me talking and my garrulous daughter giggling. The conversations are mostly like FAQs:

Qs: How is the Wi-Fi network today?

Ans: Working well in good range.

              (Wives, of course, work an even wider range.)

Qs (to son): Have you bathed in the last 48 hours?

Ans: Hmmm....

Qs: Have you received your rota of household chores on the family WhatsApp group?

Ans: Hmmm....

              Yes, I created a family WhatsApp group; it saves energy, eliminating the need for verbal reprimands. And we stay in touch too. We also post pics – clothes left in the bathroom, milk spilled while boiling, food

UBUNTU: I AM BECAUSE WE ARE

dropped on clothes – picturesque reminders of each others' flaws. Besides, they act as proof when one of us is in denial. *I didn't do it. Oh really?*

Qs: Did you remember to give the maid a stock of fresh masks?

We do not check our maid's temperature and saturation as our neighbours do. But her absence does make me feel hypoxic and breathless. Mind you, using these screening tools – pulse ox meter, infrared thermometer, UV disinfectors, in housing societies and homes seems to have become the new normal status symbol in the Covid era. The more elite one is, the more sophisticated the tools. Some even have mini robots to sweep and mop. In my house, however, I cut old towels into four and use them as mops. I have not yet got to the stage of cutting up my husband's vests though.

Love & Monsoon Mania in Covid Times

Hopeless romantics usually get all fuzzy with the onset of the first rains. Some even start listening to Bollywood monsoon songs and sigh about taking long drives to Lonavla, in the waterfall draped Western Ghats.

For the 40 plus or the less romantic of heart, the rains bring arguments about who will dry the clothes, get the groceries, or even fry the hot *pakodas*. Being a quixotic soul, I told my better half the other day that I wanted to go for a jog in the rain. His response was: 'It's pouring cats and dogs. If you go out, you will get sinusitis, slip disc, and knee joint swelling or Covid-19.' Sigh. I wished I had a more romantic reason to feel my knees go weak or a shiver run down my spine.

Bollywood Romance in Covid Times

In the black and white Bollywood era, sizzling love scenes were depicted in a subtle manner, with two flowers touching or horses nuzzling. In the Covid era, it might perhaps be depicted as two colourful masks swinging in the wind and touching, or mask strings getting knotted.

Ashiqui could be remade for the Covid era with the famous umbrella scene having the couple in

PPE suits and kissing through their masks under the *de rigeur* umbrella. *1942 A Love Story* could be remade as *2020 A Covid Story.* The French kiss might well become history, replaced by the Wuhan kiss.

Marriages are made not in heaven but online (the new normal)

The Big Fat Indian wedding has shrunk for good, with the bare minimum of people and the priest reciting *mantras* on screen. *No frills, only thrills* is the new motto for Covi-dial couples getting hitched. Even hostels and hospitals are now accepted marriage venues. Quite novel and exciting, don't you think?

Covid has created a pause in our lives. We all wish this was a surreal nightmare, but alas it is not. We all crave, with hope in our hearts, to return to normal – to meeting friends, eating in restaurants, going to school, and playing football. We have a new appreciation for freedom.

Fear and uncertainty have challenged our wits, but also re-jigged integral aspects of life, like love and compassion. It is *love* when we instinctively think to protect our elderly parents or children from infection. It is *love* when, as a team, we share household chores as a family. It is *love* when you save a piece of the cake you have baked and sneak out in lockdown to share it with your best friend. It is *love* when your spouse fixes you a steaming *cuppa chai* when you have had a tiring day. It is *love* when you take in snacks for the overworked Residents and nursing staff at the hospital.

The times ahead may get even tougher and social distancing may prevail as the future norm, but we can find ways to bridge artificial barriers. Let us be kinder, more tolerant and compassionate. Let us help those less fortunate, those who have little or nothing. Let us be more accommodating and overlook small flaws. Let us respect others' likes and dislikes.

Love conquers all. I am confident that in the end, love for humanity will prevail and we will emerge as Shiny Happy People. *Ubuntu.*

E-GURU-KOOL

Education is not the beginning of facts but the training of the mind to think. ~ Albert Einstein

'Mom, wake me at 8am. Can't be late for e-classes,' chirruped my daughter. When you're a mother, you're not just a person, but also an alarm clock, dictionary, Google, computer, robot, and Alexa. And the motherboard can never crash. 'Why not shake up the world and ask your father for a change?' I suggest. 'No, Mom,' comes the answer, 'I don't like the way he wakes me up. He just shouts and flings open the drapes, unlike you waking me with a loving cuddle.' I sigh and wish the Covid 19 virus would manifest some feminine traits in men's brains as well. The world could then be a more gender equal place AC (After Covid).

From sitting under a tree in *gurukuls* to AC classrooms, education has taken a leap of luxury, creating holes in parental pockets. But who would have thought that one day school would come home instead of the child going to school?

I had always imagined home-schooling to be the domain of celebrity kids, or perhaps eccentric parents.

There are certainly some benefits. Moms are no longer required to cudgel their tired brains about packing tiffin at first light, or drive in a frenzied state to drop off children at school before dropping themselves off at work. Bliss.

But there are challenges too. In my house, in one room, my daughter's school is in full e-action; in another room my son is online with his med-college. I run coffee and breakfast service between school and college. The kids' dress codes are doozies. My daughter has on her school shirt with I-card, and shorts, while my son often brushes his teeth during the first lecture – Anatomy.

Like a typical Indian mom, I find reasons to sneak peek into both academic halls to make sure neither is offline and snoozing. With the kids at home, there is no longer any ME time for moms (fathers please note). Parents are also required to pitch in to monitor homework, help complete projects, and attend online PTA meetings. If this goes on, it is entirely possible that it will be a parent who comes up with a Covid vaccine for children.

But the task of teachers is even more daunting. Taking attendance in 2025 could possibly include names like Covid Awasthi, Quarantina Joshi, Social Distance Singh, Mask Mehta, Atmanirbhar Kelawala, Sanitizer Shrivastav, Pandemic Pandey, and Vaccine Malhotra. For now, they have to deal with latecomers in night pyjamas and the excuse that the laptop camera is not working. The smart generation also have the ready excuse of network issues. Some children disappear between classes to go find a snack in the kitchen. In India, blame it on Wi-Fi or blame it on the Wife.

Then there are those awkward moments and situations teachers have to face. They cannot 'unsee' towel-clad Dads passing in the background, Moms in nighties, little siblings running naked, and even guest appearances by pets. They are treated to the home audio as well – the ranting, the nagging, pets barking, Grandpa complaining, Mom and Dad disagreeing, the phone ringing... Distractions and disruptions abound. But the teacher can no longer punish the errant ones by saying they have to stay back and do detention or work off demerits. There was once a time when no child carried a phone. Ironically, today school is on the cell phone.

But are children happy with this new normal e-learning? I asked both my children this and without hesitation they unanimously said no. They miss the face-to-face interaction with teachers, the liveliness, the energy and disagreements with friends. If there is one thing that virtual media has robbed us all of, it is human contact — the warmth and affection you receive from your teacher, the pat on the back when you perform well, the reassuring arm on your shoulder when you do not, the hugs and hi-fives with classmates, the camaraderie of inter-school matches, the feeling of belonging to a school community.

The pandemic has caused shutdown of schools and colleges across the globe. For the first time in history, education has been disrupted in this way. 1.2 billion children globally are out of the classroom. Perhaps it is also a time to ponder on the system. Has our education system become too tied to rote learning, become too competitive? Has it robbed our children of the ability to think critically and hampered their adaptability? Will e-learning be a catalyst for change, towards more effective, flexible education which will nurture creativity and enable self-reliance rather than spoon-feeding? Perhaps the greatest learning experiment in history, triggered by the pandemic, might prove to also be the beginning of a new and better education system.

The Government too, has realised the need for a revamped education system. With major changes in degree course content, structure and tenure, we can hope to shed rote learning. With greater inclusion and more emphasis on experiential learning, we can hope to move towards true, lifelong learning.

But what of the thousands who cannot avail of mobile phones, who do not possess laptops or access to digital platforms? The poor and disadvantaged suffer the most, as always. And there are no quick answers.

HAIR RAISING TALES

Rapunzel, Rapunzel, let down your hair.
~ From Rapunzel, the Fairytale

The other day I was waiting in queue to buy groceries. The man ahead of me turned out to be a colleague. I had not recognised him with his mask and overgrown hair. He now looked more like a rock singer than a Paediatrician. I apologised for not recognising him and we had a good laugh over it.

The shutdown of salons and parlours due to Covid has made bad hair days the new normal. While men sport long neck-length tresses and women facial hair, we can now say with some truth that gender disparity belongs to a bygone era. Feminists can breathe a sigh of relief as both sexes have not only begun to look alike, but behave in the same way. Men have taken to household chores and experience MMS (male mood swings), as well as withdrawal symptoms from the abrupt termination of BNO (boys' night out).

UBUNTU: I AM BECAUSE WE ARE

My cousin called me one morning to say her husband, in an early morning groggy state, had actually confused her with someone else because of her suddenly grey hair. The pandemic has certainly added years to our age, not because of stress, but because we finally dare to bare the grey, thanks to the shutdown of salons.

With the silvers standing out in my hair too, I was lamenting this to my better half when he reprimanded me saying, 'Why can't you stop colouring now and age gracefully?' He had a point. Nevertheless, I quipped that I did not see any 'silver lining' in this silver.

When he complained of an itchy scalp and overgrown hair, I offered him my 'barbaric' services, using kitchen scissors. After watching a few You Tube videos, I felt more confident, though my skill remained questionable. While most men face with true martyrdom the razor sharp tongues of their wives, they are now compelled to also suffer a DIY assault by their wives on that most prized male possession – his head of hair.

Perhaps love is in the h-air? Actor Anouska Sharma trimming cricketer husband Virat Kohli's hair probably inspired many. On the other end of the hair scale, quaran-teens seem to prefer to look like apes. My son's appearance rivals posters of Mangal Pandey with his uncombed flowing locks. But he refuses to risk his mother's barber skills and kitchen scissors.

Bald people are perhaps the ones having the last laugh. And many have taken a leaf from the Prime Minister's look and grown long beards to match their neck-tingling hair. Actor Tom Hanks has nothing on them in *Castaway*. Some celebrities even flaunt their hairy progress on social media.

Women have their own woes. Smooth and silky is no longer the in thing, but raw and real. *Feel it or deal it* could perhaps be the new tagline for hair removal creams. Visiting the parlour is therapeutic for most women, like seeing a shrink. It's a place to chat while getting pampered. I am something of an aberration in this regard as I tend to have panic attacks if I spend more than 30 minutes in such places, and prefer my ME time at home. Perhaps we will see a Bill introduced in Parliament soon, to make salons and saloons part of essential services.

I now wonder if Rapunzel was the perfect example of home quarantine and social distancing in her tower. And since her Prince did indeed find her

UBUNTU: I AM BECAUSE WE ARE

even there, there is perhaps hope for all the single women in lockdown.

Unable to bear the weight of my uncut hair, I called my hairdresser to ask if he would consider a home visit, like my patients ask of me. After some initial reluctance, he agreed. I offered him a family package – three generations of heads, creating family history. Later, I even collected some locks in a zip-lock bag as a Covid souvenir.

So, while he was trimming my split ends, I tried to understand the challenges he had faced in lockdown. He told me it had been difficult financially, emotionally and mentally. He got daily phone calls from clients but could not risk his family by working. At the same time, the rent on the parlour and staff salaries had to be paid. And once the parlours opened, there was the added cost of sanitisers, masks, coveralls, etc. It was a big financial burden. Clients naturally enquired into every aspect for their own safety. But what of the safety of the staff? Many clients casually unmask while getting a haircut, shave or facial. Many refuse to pay the extra charge for the coveralls and disposable linens used.

Our chat brought home to me the reality of being humane, of being considerate of the safety and livelihood of others. It is so easy to sport a T-shirt with a pious tagline, but it is time our actions speak instead. The pandemic has robbed us of many things, both big and small. But let us remember we are in this storm together, in the same boat. While the fortunate can sit in their cozy homes and watch Netflix, there are many who have lost jobs and incomes and cannot make ends meet, pay the rent or their children's school fees.

So the next time you visit your salon or parlour, let your hair down but not your mask. Ask about those who serve you. Pay the extra. We can come out of this with flying colours and perhaps even sporting a covido-fringe.

DAMSEL IN DISTRESS

I wanna be free, like the bluebirds flying by me,
like the waves out on the blue sea...
~ Song written by Bobby Hart & Tommy Boyce

March 8th, 2020 BC (Before Corona): Women all over the world celebrated International Women's Day, claiming their freedom and basking in ME time parties and events.

25th March 2020 AC (After Corona): Lockdown and the start of Wo(e)men Woes.

Shanta, my house help, was an intrinsic part of my daily life and routine. With Shanta gone during lockdown, my life was *ashant* (no peace). At least the Chinese had robots.

Who said the lockdown brought us closer to nature and gave us time to 'stand and stare'? William Henry Davis was, after all, a male. I just stand and stare at the heap of clothes, the toilets that need a scrub, the dust on the furniture, the couch where my two big potatoes (children) sit till mid-day. The interval between their waking and sleeping is exclusively my time they say. 'It is your me-time mom.' Me time? What's that? I am the house help, chef, baker, dhobi, cleaner and shopper all rolled in one. The Complete Woman, in fact.

Agreed, there are no tiffins to be packed for school/college/office, but then aren't packed lunches easier than three course meals for four? I feel I am on a culinary crusade and the family on an odyssey of gluttony. Since they don't have anything to do, all they need is food. I would happily do two hourly NICU/PICU rounds rather than two hourly cooking and feeding cycles at home. *Ma ka Dhaba* (Mom's Kitchen) open 24X7. Emulating the Covid virus that visits each country, the kids demand Italian, then Mexican, then Chinese. To top all this, there are the superwomen who keep posting exotic dishes on social media, which seem to trigger the food centre of the family brain. My hands feel like sandpaper and smell of onion and garlic. My clothes smell of disinfectant from sweeping and swabbing. I have even picked up the maid lingo.

When my Resident called one night, I almost prescribed a 'teaspoon' of Allegra, as if it was baking soda. I am worried that if this continues, I might forget paediatrics and read only cookbooks.

The nadir arrived the other day when I found myself watching a YouTube video of how to get rid of garlic smell from your hands.

And hips don't lie. The anthropometric diameter (mine), and waist-hip ratio, have been inching past the 90th centile since gyms were locked down. I have gone from Fabulous to Flabulous, and need to Flab-U-Loss soon.

Then there is the overtime with the boss (spouse). Don't remember spending so much time together even on our honeymoon. We are clearly 'maid' for each other. He cleans the mess I make in the kitchen and tries to chop vegetables as meticulously as he excises cataracts from his patients' eyes. The other day he started leafing through old photo albums murmuring, 'You have changed so much.' That did it! The account summary of the last two decades lasted the whole day. It began to feel it would be easier to break the Covid chain than this quarrel.

As if all this was not enough, the kids' online classes, projects etc on Zoom make me feel giddy with stress vertigo. While there are apparently those in the world who are competing for 'Survival of the Laziest' award, I remember all the women who will need seven days leave post lockdown just to recover from the fatigue and reboot ourselves.

I truly miss some important people in my life. No doubt absence has made the heart grow fonder.

Dhobi: '   *Bhabhi*, keep your ironing out.' Bliss.

Milkman: My wake up call. How I wish it was a *chaiwala* (tea boy) instead, delivering *cuppas* at 6am.

Maid: Our lifeline, our oxygen. I actually dream of the sunny day when she will ring my doorbell and park her *chappals* outside, saying, 'Didi, I am back'.

The return of our house help will be a historical moment in the lives of women post Covid. I'm sure we would happily clap and bang utensils in their honour, as we did to express our solidarity against Covid. Perhaps the Prime Minister will consider this tribute to these truly essential workers.

With all this juggling, there is so much emotional upheaval as well – CMS (Corona Mental Syndrome) – irritation that there is so much to do, exhaustion, impatience with the lockdown, anxiety that everyone must be safe and healthy,

UBUNTU: I AM BECAUSE WE ARE

and the longing to get back to normal life. I am
sure every woman caught up in the turmoil of
our Covid-induced lives has only one real wish –
to be free.

ATMANIRBHAR

Expect more from yourself than from others for in the end all I learned was how to be strong alone.
~ Anon

The economic crisis triggered by the Covid pandemic gave birth to a very important national campaign – *Atmanirbhar Bharat Abhiyaan* (Self-Reliance). It was announced by the Prime Minister in May 2020. Though it may seem a tad emulative of the Swadeshi movement during the British era, with many calling it old wine in a new bottle, it nevertheless lifted our spirits.

Other taglines like 'Vocal for Local' or 'Make in India' underscored the point. However, self-reliance does not mean cutting oneself off from the rest of the world. We have perhaps been over-dependant on Made in China. 2020 taught us how dangerous such reliance is. The border skirmishes with China, in which many soldiers lost their lives, only served to highlight the growing need to be self-reliant and strong. There followed bans on many Chinese products, apps and solutions, encouraging homegrown entrepreneurs to come forward and plunge into creative and innovative manufacturing. A new era had arrived, unsought but real.

India's self-reliance will be based on the five pillars of EITDD: Economy, Infrastructure, Technology, Demography, and Demand. But we Indians have taken the appeal so seriously that we have applied this new *mantra* to our personal lives as well. Schools have made it a part of their curricula. So much so that competitions are held on the topic, essays and poetry written about it, and plays enacted on the theme. Even a kindergarten child today is familiar with *atmanirbhar*, no matter that they might struggle with the pronunciation.

My family too, has taken it seriously. My better half is an ardent fan of the Prime Minister, so his clarion call has been religiously implemented in our home. As the disciplinarian parent, he announced that both our children would do their dishes, make their beds, clean their cupboards etc, and be self-reliant on the homefront. But the Prime Minister, being a man, perhaps failed to realise that we women are already self-reliant and have been multi-tasking at home and at work.

I did announce to the family that each of them would take turns to cook the meals so that I got a break. Alas, the *mantra* broke down at that point. When I declined to make my son's coffee, he retaliated by telling me to be *atmanirbhar* when I asked him to go and get some groceries. My daughter innocently asked if Alexa and Google fell within the scheme. When you next ask Alexa to play your favourite song, she may tell you to sing it yourself.

Perhaps singles are the happiest with this new scheme for they have been self-reliant in all spheres anyway. For the rest of us, we must learn new ways. For example, if you feel deprived of love in your life, just say ten times a day: I love me. Why depend for love on others? New Bollywood movies will depict Raj telling Simran to get to the train on her own. Young girls will henceforth tell their parents they will find their own life partners. Mukesh Ambani, the Industrialist, must of course, be smiling broadly because he *is* Reliance.

Whether our country rises to this challenge with outstanding outcomes only time will tell. From zero PPE production in March 2020, to the production of 2 lakh PPE kits daily, India has certainly demonstrated the spirit of self-reliance and the ability to see opportunity in adversity. And the repurposing of various automobile plants to collaborate in the making of lifesaving ventilators is another step on the thousand-mile journey towards resurgence of our national economy.

Unlock guidelines have enabled a gradual resumption of economic activity, while maintaining abundant caution. The rest is up to each individual: wear masks, maintain social distancing, wash hands, sanitise premises, do not crowd...

Atmanirbhar India can have a transformative impact on a newly empowered, self-reliant, young and modern India. But the road is long and full of hurdles to overcome. But *Make in India* is the song on everyone's lips, and *Self Reliance* our new motto.

FOOD PANDA

If you can't feed hundreds, then feed one.
~ Mother Teresa

It is rightly said you are what you eat. The Chinese ate bats and the rest is history. We all wish they had eaten apples instead. Of the many indulgences the pandemic deprived us of was gastronomic pleasures. Be it *pani puri* at a roadside stall, hot and spicy *vada pav* or a candlelit exotic meal at a five star restaurant, it all seems to have become scenes from a sepia toned movie of yesteryear.

With the lockdown we all adapted to simple staple food as in the ancient Harappan era. Dal-rice, *khichdi* and pulses. Since many feared to eat vegetables contaminated with the virus, healthy food took priority over appeasing the tastebuds. Hardcore carnivores turned herbivores, not to save the planet, but themselves. Chicken disappeared from the menu along with Bird Flu.

But humans are resilient creatures and food cravings cannot long be disregarded. Our palates got bored of the same platter after a few weeks of eating healthy, like a Panda bored with eating only bamboo shoots. Irresistibly, our epicurean desires resurfaced. My children craved Mac D, Dominos, Belgian Waffles...the list was endless.

Meanwhile, many women took up the gourmet challenge seriously and switched to the Masterchef role. Men too, dared to step into the alien territory of the kitchen, learning new skills to kill time or just to be in their wives' good books, only to discover a new and fascinating world of flavour and fun.

Social media was flooded with culinary showpieces and luscious images of dark temptations – cakes, desserts, puddings, Dalgona coffee. Even *thali* pictures were posted. Recipes were shared on WhatsApp groups and a few posted You Tube videos too. Meanwhile, I was serving spikey (aka Covid) *sabudana wada* at home. I banned Chinese food, much to my children's dismay. When they rebelled, saying noodles were their *hakka* (right), I cooked Malwani Manchurian in coconut gravy and Surti noodles Gujju style.

Many of us either binged like gluttons or indulged in stress eating, consuming more food in response to emotional stress. Just as the Panda eats for 12 hours, quarantined people ate the whole day long. No wonder they call it a pandemic. The result was that our waists expanded with the Covid curve. To the extent that the buttons of my jeans socially distanced themselves and would no longer fit my flabulous waist.

The pandemic brought us to our knees. Businesses and aspirations deflated in deadly dualism. The restaurant industry was hit particularly hard. With eateries shut, staff retreated homeward, many without any foreseeable income, many jobless. Ironically, those who had once served hungry clients were the first to be left hungry. According to the National Restaurant Association India, the industry suffered a loss of approximately INR 4,23,865 crores, with 77,00,000 employees affected. Even in the unlocking phase, when Swiggy and Zomato initiated safe deliveries at the doorstep from hygienic kitchens, most people still feared to eat outside food. Though the virus is not transmitted through food, apprehensions about hygiene and en route transmission lingered in people's minds for a long time.

So what will the future of dining be? Will social dining entail sipping *quarantinis* inside our homes? Or will it be like a science fiction *Matrix*

UBUNTU: I AM BECAUSE WE ARE

style movie, with restaurant seating featuring screens between tables, menus read on gadgets, staff in PPEs, and the food arriving from cloud kitchens? Will our meals be seasoned with a variety of immunity booster toppings instead of cheese and paprika? And when we are done, will drones sanitise the table for the next guests?

As the science writer Ed Yong wrote, we must stop thinking of the lockdown as a blip after which life will go back to normal. Instead, we must prepare to find a new normal. Long term changes will have to be made about how we live and behave.

The future might be dark but there is always light to be found at the end of the tunnel. A bunch of my old friends decided to meet up as in the good old days. We met outside one of the small eateries we haunted in our college days. We spread the hot *idlis* and *pav bhaji* we had brought, buffet style, on the car bonnet. The hot *sambhar* satisfied the palate and the company satisfied the soul. For us Indians, dining is not just about eating, but about bonding and having a good time.

While the fortunate have enough food on their plates, there are millions who face hunger, trying to get one meal a day. Fortunately for the world, humanity is still alive in humans. People and organisations of all hues have come forward in a truly altruistic spirit to provide meals and tiffins to those in need. Perhaps we can all be part of this small act of giving, what I call the 'kindness plate' – a plate of food cooked with love, to be passed on to someone hungry or needy. If we can all do something for someone, however small, imagine the joy we can spread along the way.

We are all caught in this strange limbo called Covid, and it is often hard to imagine a world beyond it, or find hope and joy in the midst of so much uncertainty. But what we can do is feed a hungry stomach, which will in turn feed our hungry hearts.

TO KNOT OR NOT-TO-BE

Wise men say only fools rush in
but I can't help falling in love.
~ Song written by Hugo Peretti, Luigi Creatore
& George David Weiss

They took their wedding vows and he kissed her under the moonlit sky in the presence of the many assembled... Thinking back to weddings as they used to be celebrated once upon a time in BC (Before Corona), has the nostalgic aura of sepia tinted memories. Many couples who had eagerly planned their big fat Indian weddings months before, were hopelessly stalled by the extended lockdown. Some postponed the event to the next year in the hope that the vaccine would end the pandemic by then.

But an intrepid few, eager to eat the wedding *ladoo* (the aftertaste of which lasts a lifetime), revamped their wedding plans to fit the new Covid style. These digital and intimate events sacrificed the hoopla and overeating. They became Zoom events, with the guests dancing in their own homes and showering blessings at the screen. It was certainly 'an affair to remember', to be narrated to grandchildren in years to come. But lockdown could not prevent locking these couples in a lifetime of togetherness.

Every bride and groom dreams about their wedding day. Parents too, have their own aspirations for their children. Perhaps a big procession, fire crackers, 500 invitees, a *sangeet*, a *mehendi*, exotic cuisines, make-up artists, photographers... But all this has taken a hit during the pandemic. New norms of not more than 50 guests make wedding-planning a perplexing affair. Who does one include? The situation is fraught with danger. The ones who attend are masked (entailing perfectly matched masks for the ladies), and offered sanitisers instead of perfumed water. The installation of sanitisation fans in place of mist fans is now the vogue. And one cannot forget the *de riguer* temperature thermo scans for all attendees.

Venues have shifted from lawns and beaches to living rooms and terraces. Designer bridal ensembles have been replaced with the bride wearing her mother's vintage sari, and new jewellery substituted with her grandmother's heirlooms. Necessity rather than nostalgia drive

UBUNTU: I AM BECAUSE WE ARE

these choices, but brides still make the most of what they have, focussing on eye make-up and hair as the only visible attributes above the mask. The groom is thus reminded of his favourite film, *For Your Eyes Only*, while his lips remain sealed behind the mask. Wise men soon master the art of being the silent partner at home. The ideal husband learns to understand every word his wife does or does not say.

Everything is done according to tradition, though virtually. *Pundits* too, have become tech savvy and recite the *mantras* on screen as the couple take the *pheras* in their own homes, with only close family in attendance. Kodak moments are captured by relatives on iPhones.

Being Indian, food remains top of mind. However, the fare is home-cooked, healthy, hygienic food, or pot luck. A few go a step further to include an immunity boosting option – *herbal kadha* instead of alcohol, *haldi dudh* instead of *kheer* and puddings, and probiotic yoghurt, avocado with honey and fresh citrus fruits to replenish Vitamin C. Perhaps a packet of pills – Zinc, Vitamin D3 and Vitamin C is next. But a definite benefit to this pared down wedding repast is that everyone can actually sit down and eat instead of balancing a plate on three fingers. How wonderful!

In essence, the new normal begins to sound similar to weddings of ancient times, when simplicity was the ultimate sophistication. I wonder if we are redefining our ideas of luxury? And there is so much less wastage of food and saving of money, both of which can be donated, or invested for the couple's future.

What we witnessed in lockdown was that it is entirely possible to retain the essence of romance and the gravitas of long term commitment with utmost simplicity and poise, sans *band, baja, baarat.* No frills is the new Covid norm in solemnising weddings. But the vows remain the same.

Whether they live happily ever after is history to be recorded down the line. Each partner secretly believes in after years that the other got the better deal. Marriage is like having only one channel on television, yet you still want to watch it. After all, love recognises no barrier. It even overcomes pandemics and lockdowns to arrive at its destination, full of hope and dreams.

ROLE REVERSAL

I am proud to be who I am –
a doctor to my last breath.
~ Author

We physicians may be considered Masters of Medicine and experts in treating the ailments of the human body, but the truth is we have no god-given armour, like Karna in the *Mahabharata,* to protect us from disease on the battlefield. Hence, the warrior often becomes the victim. There are times when the physician must face the reality that s/he is no longer the one wearing the crisp white coat but the one in the hospital gown; that the tables have turned.

Primum non nocere (First do no harm), is ingrained in our brains as the core principle of our medical profession. Because of it, we are chary of making life-threatening mistakes while treating our patients. This principle exponentially increases when two physicians come together, but one is the patient. Doctors do not make good patients. With them there is never the balm of blissful ignorance. And we often face another symptom – the VIP entitlement syndrome.

Doctors have justly been recognised as Covid warriors in this pandemic. But, despite taking all precautions, many of us did contract Covid 19. While the fortunate ones had only mild symptoms, others had a stormy ICU experience, but were lucky to escape the final tragedy. It is an undeniable professional hazard. Nevertheless, it brings tears to my eyes to see young resident doctors succumbing to death, their dreams and aspirations cut off before they can even take wing, or the many bereaved families of colleagues left to fend for themselves.

When a friend of mine complained of fever and bodyache, we both knew it could be Covid, but hoped it was not. Knowledge makes doctors bad patients. When his fever continued for the third day, we decided to do a Covid RT PCR test. Meanwhile, being doctors, we had already played in our minds all possible scenarios and complications (with the worst case scenario first in sequence), treatment protocols, hospital options, and in case a ventilator was needed, what the settings should be. Also, which drug or cocktail of drugs should be administered.

UBUNTU: I AM BECAUSE WE ARE

The worst part in my friend's case was the 24 hour wait for the reports. The suspense was killing. I now understand the plight of our patients awaiting biopsy reports for cancer. Finally, when the report did arrive, it was positive for Covid-19. He was immediately admitted to a hospital, lucky to get a bed and timely treatment. He recovered well. But the whole experience of being a patient changed his perceptions, and mine, giving us a clearer understanding of our patients.

As physicians we are used to being in control, always inking prescriptions and giving advice authoritatively. But when you are the patient, you feel powerless. It is like Spiderman losing his web. Your body now belongs to others; nurses take your TPR, vital parameters, intake-output, secure IV lines; you are shipped all over the hospital for tests and scans in those unflattering hospital gowns. The technicians instruct you, junior doctors do examinations. I remember when I underwent spine surgery a few years ago and the nurse came in with a bedpan to pee in. I had never felt so vulnerable as in that moment.

Covid causes loss of smell and taste, as well as appetite. But your doctor insists you eat well to rebuild your strength. *How?* you wonder wearily.

The Covid ward is different from others. You are isolated and alone (in fact, you get a lifetime dose of me time). There is minimal human contact and one feels like an untouchable. All those who enter are clad in PPE suits and shields, and look the same, like visiting creatures from space. It is like being in a spaceship; you don't know where you will land. Your mind is fogged with fear and insecurity, your body burning with fever and aching to the bone. Yet you must force yourself to eat and drink. Mobile phones have proved to be the true saviours for they allow you to stay connected to the outside world. Was there a time when mobile phones did not exist?

I asked my friend whether he was able to go out of his room for a stroll. 'It's like Neil Armstrong walking on the moon,' he said. I advised him to mingle with the other Covid sufferers in adjacent rooms to share their common woes. A social creature like myself would certainly have bonded and made a few Covid buddies.

But I realised something very important. As doctors, we often fail to see the problem through the patient's eyes, or understand their perspective. Greater compassion, a gentle pat, a reassuring smile are effective in treating patients. We need

UBUNTU: I AM BECAUSE WE ARE

to constantly remind ourselves that our patients and their families place their most precious asset in our hands. We must respect their trust and recognise the frailty of our own human-ness.

My friend was discharged from captivity after a week and came home wearing a T-shirt given to him by the hospital. I thought it an innovative idea and pictured him stepping out in a T-shirt which said, *I am a Covid Conqueror*, his belly flat from weight loss, a smile on his face. It must have felt so heartening to actually step into daylight and see people on the streets, and hear the sounds of traffic. His tryst with society was, however, a brief one (the car ride from hospital to home), for he was required to isolate again at home for the next few days. So there he was, room alone, with an overdose of screen time, eating and sleeping.

Isolation is emotionally and mentally tough to sustain. Humans are by nature social animals. It is a 14-day exile from the rest of the world. In my friend's place, I would no doubt have written blogs and stories on the walls of my room in frustration. One needs mental alchemy to stay sane and resilient – with music, books and movies you enjoy, meditation, and regular shots of laughter from friends. They act like balm on an agitated mind. For once, you yearn for something negative – the Covid report.

Viktor Frankl, the famous Austrian neurologist and psychiatrist who was imprisoned in the Auschwitz concentration camp during World War II, wrote his famous book, *Man's Search For Meaning*, during his time in Auschwitz. He also invented logo therapy, which later became one of the crucial tools in treating mental illness. I would tell every Covid patient that while we cannot control the environment, we can control how we face it.

Covid pushed many doctors from wellness to illnesses. Some experienced first-hand the difficulties of getting beds, the filthy toilets, the unavailability of senior consultants, the huge costs of admission and drugs, and a few lost the battle as well. Will this spur medical associations to take up these issues? There are beds reserved for politicians, the police, and specific communities, but what of doctors? What of their families, left to fend for themselves in a void of financial instability? Neither the Government nor any NGO comes to their aid. No one thinks doctors need help.

When a soldier goes to war, that war does not physically enter his house and touch his family.

When doctors work in Covid wards or OPDs, they know they are endangering their families as well. Yet, doctors continue to do their jobs, with a prayer in their hearts for the safety of their families. When a soldier dies, his family is the responsibility of the State (perhaps not always fulfilled as one would wish), unlike when a Doctor succumbs on duty. He is neither pensioned, nor is his family compensated for doing his duty or glorified as a martyr.

While doctors certainly do not desire 21 gun salutes or their bodies to be wrapped in the Tricolour, they do expect respect and dignity. Not assaults, physical and mental, or being labelled greedy and avaricious. Consultation fees are a doctor's livelihood. The Government does not bear the cost of our children's health and education, our electricity bills, nor the hospital expenses of ageing parents. Nor do we have Government pensions or financial aid for spouses on our demise.

But I do believe that what you do comes back to you. That my colleagues have survived, as my friend did, is because of good deeds done unknowingly as doctors. I would also salute all the nurses who work long hours in PPEs, taking multiple shifts, the junior resident doctors who bravely and consistently remain on the battlefield, and my doctor colleagues who work with humility and humanity. It is wonderful that the best traits in humans live on, strong and well. The only reward doctors want is for their families to be safe and provided for, a word of appreciation from those we serve, and some patience and understanding if you are not immediately attended to. There are others like you.

NON-INFECTIOUS FESTIVE SPIRIT

Blessed is the season which engages the whole world in a conspiracy of love. ~ Hamilton Wright Mabie

The virus has cov-ertly altered the normalcy of our lives as the SARS saga continues. The Big C is now Covid. Even my house help has a new mantra – sanitise. Nowhere is this disruption seen more plainly than in the calendar of Indian festivals. Our Covi-dieties did indeed arrive on their annual vacations (visits) to Earth, but we humans were all home on staycation. *Ganpati Bappa* arrived on Planet Earth and then retreated, asymptomatic. *Durga Ma* stayed for nine days and then returned to her heavenly abode, also asymptomatic.

Meanwhile, the Covid virus seemed to have taken up permanent residence. Is it our hospitality it likes or the bio-war with humans? *Navratri* came, with people offering prayers at home or digitally to the nine avatars of Devi, but the virus took it as a sign to mutate into new avatars as well.

There was no festive fervour. The glitter and excitement of Indian festivals had turned to monochrome. Streets were deserted and shops empty. People did not have the money to splurge on clothes and jewellery. Purse strings were pulled tight. There was no gusto to dress up since there were no social gatherings or visiting *pandals*. God had gone digital, arriving with mobile phones as their *vahanas*. *Aartis* were on zoom, and *prasadam* offered virtually.

I felt for the Bengalis during *Durga Puja*. How would they weather a *Pujo* without crisp new Tangail sarees and *bhog*? After all, *coveed* or *shobheed*, *sindoor khela* is their birthright, just as crowding the pavements of Gariahat is. Similarly, the loquacious Gujarati community might have to wear *tope class* designer PPEs, but they will still do *gharma garba* and sing *Eh halo halo*. Falguni Pathak, the *garba* doyenne, has been booked solid for virtual performances, I hear. And not to forget the *Dassera* snack of *jalebis* and *fafda*. Can there be paradise on earth without them?

Meanwhile, inside homes, there has been an overdose of togetherness and spouses are bursting plenty of fireworks. Wives have taken the *Mahakali avatar* and men that of *Ravana*. Children are fighting like the Kauravas and Pandavas. If we emulated Greek mythology, the

UBUNTU: I AM BECAUSE WE ARE

wives would at least look like Cleopatra and the men don shades of Mark Antony.

Diwali was a noiseless, dark affair. Our Prime Minister urged, 'be your own light'. Diwali cleaning was more a sanitising effort. Women prayed that Laxmiji would understand and enter their homes regardless of Covid. No Diwali bonuses were forthcoming and there was a general pall on spending. The indulgence of dry fruit boxes was replaced with peanuts, and gifts of sanitiser.

Festivals revive and reinforce culture and social traditions. Togetherness with friends and family fosters camaraderie and bonding. The festive ambience kindles the spirit and gives us a break from the daily ratatouille of life. But who could have forseen the gloom of 2020-21? *Ganesh Chaturthi* without Ganesh *murthis* to worship, *Durga Puja* without *sindoor khela*. *Diwali* without lights and gifts. *Christmas* without mass and ringing bells. This is the reality we live with.

Yet there is a silver lining to everything. These are unprecedented times but we can still keep upbeat through small acts of kindness, by helping someone, buying clothes for the children of your house help, distributing sweet boxes in orphanages and old age homes. If you light a lamp for someone else, it will never fail to brighten your path also. And there is nothing stopping us from dressing up in our Kanjeevaram saris, silk kurtas and Modi jackets, making *rangolis*, lighting lamps, creating sweet indulgences, and bonding with family. Let us shed darkness, fear and negativity from our minds, and illuminate our hearts with gratitude for all that we still have amidst this pandemic.

Yes, Big C is in the air, but Big S is in the Universe – the undaunted Human Spirit. We *shall* triumph together. *Ubuntu.* Till then, keep shining and spreading good cheer. Celebrate each day of life as a festival of joy.

THE RUSH TO HANG UP ONE'S BOOTS

They also serve who only stand and wait.
~ John Milton

It was a Saturday morning and my clinic was almost empty. Three patients sat waiting, maintaining social distance. I missed the hullabaloo and noisy pandemonium which once existed – crying children, mothers exchanging baby woes, and dads patiently waiting. The silence at my work place seemed unbearable to me. To add to this, my secretary lamented that the monsoons had not brought the usual rush of cases. I remind her it is a good thing that children are healthy at home.

My better half, an Ophthalmologist, has been *eyedle* since the lockdown began. So the children and I are compelled to undergo weekly eye check-ups. While hospitals are at full occupancy with the spiking number of Covid cases keeping physicians on their toes, all elective surgical and non-Covid work has come to a halt. Many small non-Covid nursing homes are also incurring heavy losses while still bearing the outgoings of staff salaries, rentals, electricity bills and other expenditures.

I happened to meet a friend in the car park, who runs a garment shop. He said there was no business at all in the last few months. Only losses. Since the only dress code now is PPE and mask, who was interested in buying Manish Malhotra or Ritu Kumar labels? Only a few newly married couples, enjoying a prolonged lockdown-honeymoon, might indulge. For midlife couples struggling to maintain sanity at home, it was strictly essentials only.

I call the owner of a Motor Training School, who was to have my car papers submitted to the RTO. When I call him at noon, he sounds groggy, so I enquire about his health. He says he has been sleeping away the day since there was no business and does not know how to pass the day. Somnia has become an escape for many people, to beat the boredom of lockdown.

In the initial weeks of lockdown we all saw as a much needed break, a time to rediscover hobbies, catch up with family, play with children etc. The second month we still kept our morale high and hope afloat, saying the curve was sure to flatten

UBUNTU: I AM BECAUSE WE ARE

and we would resume normal life. Jokes about Industrialist Mukesh Ambani finally being able to tour his whole house thanks to the lockdown, went viral. Husbands tried to ingratiate themselves with their wives as the only eatery left open was *Bibi ka Dhaba*.

So we continued our extended home stay trying to keep ourselves cheerful with reading, eating and watching movies. And household chores kept us busy too. But after a few months of this, people began losing their patience, resilence and ability to cope. Each month was navigated with the hope that the next would be better. Alas. Our emotional thresholds were remorselessly lowered and our mental equilibrium became a challenge to maintain.

People began to crave, like a fish flapping out of water, to go back to work, back to being a doctor, lawyer, business-man, professional. I felt like pulling my hair whenever someone told me to enjoy this semi-retirement. I wanted to say I did not want to retire because I was passionate about my work, my profession. In fact, in all the trying times of my life, it has been my work that has helped me through, keeping me busy.

NDTA (no desire to do anything) can be good for a stop space, a hiatus before you go at it with gusto again; a short vacation to recharge. But this prolonged break has served to break morale and impose forced retirement upon people who need to earn a livelihood. It is compulsory detention. It is like expecting a bird to remain in a cage and never use its wings to fly. Remember what happened to the Kiwi? It forgot *how* to fly.

We all are waiting with bated breath for the vaccination drive to take effect and finally break the chain of transmission. We have had enough of being disconnected and socially distant. We are refuelled and recharged and eager to jump back into work, meet our friends, travel, get exhausted, and look forward to Sundays as fun days. We are bored with every day being a Sunday. Every day is a struggle to lift our spirits and contain our inner fears. I am tired of hearing that this too shall pass.

Yet, however dark the night, we can still see the stars. Though the dawn might bring yet another slow motion day, we need to be thankful we are here to experience it. Millions are not. This is a time of survival, so if you and your family and friends are safe and healthy, be grateful for it.

UBUNTU: I AM BECAUSE WE ARE

As far as managing moods is concerned, we must first manage the mind. Distract yourself with music, books, cooking. Keep busy. Shun negative discussions about Covid. Discuss happy events; add a dash of humour to everything. Talk with friends. This is a golden opportunity to connect with all those you normally have no time to reach. Help the needy with food, essentials and medicines. A small act of kindness will bring you certain joy.

Ask yourself what is in your control and what is not. Let go of what you cannot control. Worry just robs you of time and energy. And remember that patience pays off. Patience is not just the ability to wait, but wait with the right attitude.

We all have good days and some not so good (I never call them bad). Just embrace it all and give yourself up to positive healing. God's delays are not God denials. I am certain that post Covid, we will be busier than ever and we will break free from this forced Covid somnolence into a renewal of life.

THE RUSH TO HANG UP ONE'S BOOTS

WOBBLY WOES OF A ROCKING CHAIR

Those who respect age deserve to live to be old and to be respected themselves. ~ Samuel Richardson

It is widely thought that those worst hit by the pandemic were the migrant workers, but I consider senior citizens to have been perhaps even more badly impacted, with far less noise made about it.

A Dermatologist friend who lives in the UK, recently went through a dreadful experience. Both her elderly parents, aged 75 years, who lived in India, were diagnosed with Covid. One day, her mother, an Alzheimer's patient, did not realise her husband had fallen and was lying unconscious. A maid found him a few hours later. The neighbours were unwilling to help as the couple were Covid positive. A friend called for an ambulance and the couple were admitted to a Covid facility. The old gentleman was not doing well and the couple were told to transfer to tertiary care with an ICU facility. My friend in the UK tried to get an ICU bed, and finally contacted some of us medical school classmates. An ICU bed was managed and her parents transferred. After a few days her father improved and was shifted out of ICU.

A couple of days later, my friend was trying to call him on his mobile, but he did not pick up. Concerned and panicked, she asked us to help her find out about his condition. I called the duty doctor, who told me the patient had dropped his phone and broken it. The doctor was kind enough to allow my friend to talk to her father on his own phone. The next day, however, the old gentleman developed fever and secondary infection and died a few days later. Three of us batchmates performed the last rites for our friend's father. She could not come to India due to the lockdown. It left a scar on her mind and heart and for a long time she was unable to forgive herself.

Her mother recovered from Covid, but with no one to look after her at home, she was shifted to an old age home. There is so much to ponder about life in such a situation, where the human heart and ground reality are at cross purposes. What is in our hands and what is not? Do we not all desire to be with those we love when the end comes?

Senior citizens have maximum susceptibility to Covid 19 infection and death, due to poor immunity, co-morbidities or weakened health. They are also more vulnerable to any infections that younger members of the household who go to work might bring home. So most of the elderly have perforce been restricted to their homes for months on end. Lack of exercise causes joints to stiffen, and many suffer from loneliness, insomnia, anxiety and depression. Many of them are not tech savvy and the digital handicap has become a great barrier in Covid times. Unscrupulous hackers take advantage of such vulnerability to cheat the elderly of their savings or order products on their accounts. Harming those who cannot protect themselves to benefit themselves is perhaps the worst of human behaviour.

The elderly who live alone, whose children perhaps live far away, have no choice but to step out for essentials. Because of social distancing norms, they have to wait in queues for long hours to buy groceries and medicines, in heat and rain.

Many unfortunates die alone in hospital, as relatives and friends are not allowed into Covid facilities. And for many there is the final indignity of hasty last rites by strangers who

UBUNTU: I AM BECAUSE WE ARE

are themselves fearful and overworked. For the living, the lack of closure haunts the mind and heart for the rest of their days.

In an overstretched situation, when it comes to beds and ventilator support, policy makers at hospitals give preference to younger patients with more productive life remaining. Many of the elderly have no insurance or they have difficulty using it. For vaccinations, the elderly face challenges in registering online on the Cowin app, reaching designated centres, and then waiting for hours in long queues to get the jab. For those without kin, it has been a lonely and tough life.

Who has the answers?

Certainly we need more elderly helplines, NGOs working in this space, and better Government policies to make life a bit easier and more accessible for our elderly community. After a lifetime of working, raising families, contributing to the success of this nation in countless ways, senior citizens deserve to live their sunset years in dignity.

IN CONVERSATION WITH GOD

Dear God, if one day I lose my purpose, give me confidence that your destiny is better than anything I ever dreamed. ~Anon

Dear Bappa, so how are you feeling about coming to visit earth this year since the entire world is plagued by Covid-19? Fortunately, you do not need a visa and you have your own private transportation. As you know, international borders are closed down here, travel forbidden, and social distancing is part of our lives now. I wonder if this year's Ganesha idols will have masks? But even you would have to remove your mask to have your favourite *modak* (sweetmeat) treat. And will you sanitise your hands every time you eat?

Unlike other years, you will be spared the long queues of your devotees, loud music, and late night dance parties. Don't know if that makes you happy or sad. You will still have to pose for selfies with your devotees though, and take going viral on social media in your stride. But you will get to enjoy the stillness and serenity of our

UBUNTU: I AM BECAUSE WE ARE

planet. Possibly, we humans are responsible for the apocalypse we now face, as we forgot to live in harmony with nature and the ecosystem, and greed overtook compassion.

I prayed last year that you would visit our hospitals for you are most needed there. Every day I see doctors in masks, healthcare workers in PPE coveralls; I smell the sodium hypochlorite in the corridors and see the overfilled wards; I hear the beep of ventilator machines and see the fear in the eyes of anxious relatives. Behind the masks are a myriad emotions but I cannot reassure them with a touch of my hand. Healthcare is delivered from behind protective shields and humans seem like robots or visitors from space.

Do you worry that people expect so much more from you this year? Of course, every year you get supplications from your devotees – a good job, a new house, children passing exams, success at work... the list is no doubt endless. But this year there is only one common thing that both rich and poor want from our *Vignaharta* (Remover of Obstacles) – removing Covid from our world.

Bless us with good health, clear our minds of fear and negativity, fill our hearts with joy and compassion, and most importantly, bless us with courage and strength.

We await your coming with the same excitement and devotion as every year, for that is one thing the pandemic has not changed. Do inform your devotees to stay safe and be responsible and not crowd. May you bring happiness to all. And may we immerse all obstacles and sorrows in your Grace.

A SNEAK PEEK INTO
A PAEDIATRICIAN'S LIFE

*There can be no keener revelation of a society's soul
than the way in which it treats its children.*
~ Nelson Mandela

I'm not sure if I am a Covid warrior, but yes, definitely a Covid Specialist Paediatrician. Other viruses seem to have been 'masked' by the Big C in the air. I now have so much experience in treating Covid that I could add it as a credential to my visiting cards.

Overall, practice has taken a hit since the lockdown. I am happy if this is a reflection of children being healthier. Perhaps thanks to Indian moms having taken up the home masterchef challenge and feeding *mask(a) laga ke aloo parathas* (butter laden potato flatbreads) to their children, along with *haldi* (saffron) *kadha* (immunity-boosting drink). So kids are healthier, and perhaps heavier too.

But some do not want to step out into the sunshine, choosing to stay in their digital telecommunication bubble. From the whole family lining up to discuss symptoms, to everyone craning their necks to see how I look in a PPE suit, WhatsApp is the preferred mode of communication with doctors. First, there are the *audio symptoms,* where parents put fingers in a baby's mouth to make him/her cough so that I can get an idea of the type of cough the child has. Then there are the the *eye candy consultations,* decoding potty pictures taken on iPhones, for my eyes only. And of course, the *wholesome experience,* where the entire anatomy of the child is photographed from various angles by thorough parents and WhatsApped.

Strangely, most people seem to think digital consultations are free. The time and expertise of the physician remain the same but the fact that the consultation is online seems to create a zero-cost ideation. A few do visit my clinic (some so desperate for an outing they come dressed in designer wear and matching face mask). As soon as they enter my chamber, they take pictures and post them on Instagram # at my paediatrician's # baby crying #. I am grateful my mask, visor and PPE-scrubs hide my bedraggled look.

I am happy to see that most of those in the waiting area follow social distancing norms. But

UBUNTU: I AM BECAUSE WE ARE

there are always the exceptions. Here are a few garden varieties I have encountered:

1. Covibindass or the *Mask Kalandar* Type

A couple, with their three-year-old son enter my chamber and immediately strip off their masks to eagerly narrate the child's health history, showering me with their viral load, if any. For others, their masks keep slipping down like women's lingerie. Still others use their *duppattas* like a *nakab*, while the elites prefer itsy-bitsy-polka-dot masks which barely cover the nostrils and lips.

So now we have a system at the reception, where the staff first fix the patients' masks with tape (sealed with tape rather than a kiss, given Covid protocols). Meanwhile, my mask reads KIS M (our clinic brand, nothing more).

2. Covishielders or the *Singham* Type (those who think they just cannot get Covid)

A lady enters with her nine-year-old son. 'Doctor, you did not recognize us,' she complains. 'This is Mitarth, your old patient.' 'I'm sorry, it's difficult to recognise who is behind the mask these days,' I say apologetically. She tells me the boy has been eating like a Panda during lockdown and has become obese. He now has a fever and an upset stomach. I prescribe some medication and asked the mother to keep me posted. I also tell her we might have to run a Covid test if he did not get better. 'Doc, children don't get Covid,' she responds sharply, unwilling to believe this to be a myth. 'My son has been mingling throughout with his building friends. I feel schools too, should reopen. I just can't handle him at home. In fact, my husband got Covid last month and the poor guy was so compliant, unlike my son. He just remained in one room without making a fuss.' 'So did you do a Covid test for you and your son, and quarantine?' I enquire. 'No Doc, but I maintained social distancing. My husband slept on a mattress.'

Phew! So this was what we Indians call social distancing. But that was not the end. She told me that since her husband's symptoms were very mild, with loss of smell, they had self-diagnosed it as Covid, and did not opt for the test. 'You see, it's so cumbersome once you test positive,' she explained. 'The house is sealed and sprayed and you are locked in.' I was left speechless.

3. Covidphobics or the *Kuch Kuch Hota Hai* Type

A couple brought in their six-year-old. They were all wearing N95 masks and face shields, including the child. Also surgical gloves. With great apprehension they narrated the child's problem. When I asked them to lay him on the examination table, they took out a fancy imported sterilizer and sprayed the bed. 'We don't trust the local brands Doctor, so we got this imported one,' they explain. 'The local brands are fine,' I say, trying to be *vocal about local* as per our Prime Minister's advice.

The child had a mild cold and headache. 'Doctor, could it be Covid?' they ask. 'We are making him sniff Georgio Armani perfume two-hourly and checking his saturation level on the pulse oxymeter. He is already on Vitamin C, D3, zinc, and *kadha*,' they inform me breathlessly. 'Should we do a Covid test? We hear there is an acute shortage of beds. Is there *pre-paid booking* of beds available at any corporate hospital? Also Doctor, we heard that Remdesivir and that zumba drug (Tocilizumab) are not easily available. Should we stock some with a long expiry? We have kept some cash aside for such emergencies.' I had read about black marketing of Remdesivir but it now appeared that Covid drugs were the new blue-chip investment.

I felt exhausted just listening to their Covid preparations. It was like preparing for the Medical Common Entrance Exam. I wanted to ask that if I ever got Covid, would they loan me their pre-paid bed, Remdesivir and some cash? In return I could offer them a free consultation package.

When I got home after this beguiling Covid OPD, I happened to sneeze as I stepped into the house. My Ophthalmologist husband instantly said, 'You released 30000 droplets and now we are all at risk.' 'You could just have said Bless you!' I retaliate. 'Mom, go and take a shower immediately!' my daughter, who has a PhD in Covid thanks to the webinars she hears me speak at, advises.

As I am about to retire, my husband, who has discovered new fitness mantras in lockdown, tells me I do not exercise enough. 'Sleep early, we are going cycling tomorrow,' he announces. I have had had enough Covid psychology for the day and have no stamina left for this *cycology*. Gently, I tell him I am too tired to pedal, so why does he not cycle while I ride on the fender, like in the movies? That was the last I heard of the cycling advisory. I fell asleep, exhausted.

STORIES FROM THE TRENCHES

*How can a man understand and appreciate the time of
peace when he has never been in the trenches of life?*
~ Paul Bamikole

STORY ONE

Making breakfast, I mentally decoded the treatment plan for a child who had been admitted to the Covid PICU the night before. The Cardiologist had been unavailable and the child was in cardiac shock. There are situations in every doctor's life when you have to take risks and go by instinct.

I get a call from an acquaintance who needs an ICU Covid bed for her sister urgently. In the last month I have received 20-30 calls every day for the same reason. Most of the time I am helpless. 'There is a waiting list of 100, but I will try,' I tell her.

I contact my team and try to put her on the priority list. Half an hour later, she calls again to say her sister is breathless and her oxygen saturation 75. They do not have anyone to help them out. My heart goes out to them and I tell my team we will treat the patient in ER till we get a bed. She is thus stabilised and started on oxygen. A few hours later, a recovering patient moves out of the ICU and we are able to accommodate her.

In the meantime I try to stabilise my patient in the PICU. As I juggle patients and rounds, I get repeated phone calls and messages from another patient. When will the Consultant come, she wants to know? Why is she on such high oxygen? How many days will she be in the ICU? The calls, messages and complaints continue: The food is not good. Why were steroids given? There is too much light and noise in the ICU so cannot sleep. Why is she still on oxygen after four days?

This is a common scenario faced by almost all doctors in the past few months. When the whole nation is engulfed in the second wave, with no beds, no oxygen, scarcity of resources from test kits to vaccines, the death toll rising to the extent that crematoriums cannot cope, where families are segregated, husbands don't know the wife is dead or vice versa, and kids isolated from parents, where every day is a battle for survival, there are those fixated only on their comforts and wants.

I wish every person could spend one day in a hospital as a healthcare worker. To those patients who come gasping and when they recover, start demanding, I would like to say: Have you asked the doctor or nurse in PPE what time they last ate or even drank water? When they last slept? How hot it is to be in a PPE all day? How many weeks they have been segregated from family members? How much they risk their own safety to treat you with compassion, not just as their duty?

Pandemic Lesson 1: Human nature is such that we soon forget the bad times without learning any lessons or emerging kinder from such experiences. Patients forget that a few days ago they were struggling for life, for oxygen, waiting for an ICU bed to get timely treatment. If there is one thing the pandemic has taught us, it is that *Less is More*; that luxury is breathing fresh air without a mask; to be able to walk along the seashore; to meet and hug friends and family. We have to rise above our selfish needs to accommodate and understand others, and show kindness to each other.

In the end only three things really matter – how well you lived, how gently you lived, and how gracefully you let go of things not meant for you.

As I was getting out of my building elevator, I ran into our neighbour's son. The whole family was Covid positive and both his parents were in hospital. I felt myself boiling with rage that he had broken quarantine and was moving about like a super-spreader. He justified this saying he had to take tiffin to the hospital for his parents. But I refused to listen. Two days later, his mother passed away. I felt a great sense of guilt for having reprimanded him, for having been so annoyed. I should have heard him out.

Was I wrong to correct his actions, which endangered others? Was he wrong to go out of the house when he was an asymptomatic positive contact? Was he wrong not to abide by the rules? What could he have done? It was just the circumstances that were wrong. I should have heard him and suggested a solution – to get an RT PCR report and then going out if it was negative. Or I should have helped him find a way to get tiffin to the hospital.

Pandemic Lesson 2: We cannot judge someone's choices without understanding their reasons, the situation. Just because you do not agree does not mean you are right. There is no absolute right or wrong in life. The question is whether your actions are appropriate and inclusive. Do not jump to conclusions. Everyone is battling their own individual struggle, fragile, bruised or broken in some way. Let us be mindful of our responses.

STORY THREE

Nowadays, on any WhatsApp group, there is a tsunami of opinion, from science to politics. Everyone is suddenly an epidemiologist or a political expert. We criticize the system, blame the Government, and bring so much negativity and anger to bear. And what starts as opinion ends in a volcano of conflicts of interest, biases, preferences, arguments. Sometimes it is just an ego issue to prove ourselves right or make others agree with us.

Pandemic Lesson 3: We still compete and fight for power, position, possessions and property when half of mankind is fighting for air, for health, and survival, or battling hunger. We feel pompous seeing our names in newspapers and spend hours on social media to see how many likes and comments we got for our posts. Like the capricious weather, one minute we feel inflated when praised, and then hurt when criticized. We could instead invest more time and energy in helping the needy, senior citizens, the underprivileged, and in turn get more likes from oneself. The world right now needs compassion and acts of kindness, small and big. Our happiness does not have to depend on external factors and how others make us feel.

It is time that we quarantine bitterness and animosity and saved our relationships from going onto a ventilator. We have 24 hours a day, we are alive and healthy, as are our loved ones, so let's use our most important resource – ourselves – to help others and weather this storm together.

50 SHADES UNMASKED

*The best and most beautiful things in the
world cannot be seen or even touched.
They must be felt with the heart.*
~ Hellen Keller

Do not raise your eyebrows or get excited by the title, this is not about sizzling passion but how nature has caged us in our homes, perhaps a lesson to humanity for abusing Mother Earth, the other species of the ecosystem, for deforestation, mining, polluting seas and sky.

The Teacher of the Year award undoubtedly goes to Covid 19. It changed our perceptions across dimensions, forcing us to revisit our priorities. It also unveiled the good, bad and ugly of human nature and shifting behaviour of people. It permanently altered the experience of being a customer, a citizen, and an employee, but chiefly it taught us about being human. The ancient Hindu scriptures mention nine kinds of emotion (*navras*): Love, Laughter, Courage, Terror, Disgust, Compassion, Surprise, and Peace or Tranquillity. The pandemic unmasked them all.

Compassion, Fear, Hope and Courage: I experienced all these while working in Covid wards and the ICU. Working as a frontline Covid physician was my greatest learning. An entire spectrum of human emotion and behaviour was unveiled at the hospital. Before my Covid rounds, I got into the habit of writing my name on my PPE, to enable my patients to identify me as we all looked like creatures from outer space. But after a few weeks, my patients developed the art of voice recognition, like Alexa.

As healthcare workers, we mastered new ways to read our patients' reactions beneath the mask hooked to a ventilator – feelings of fear, despair, loneliness, but also hope and courage to fight and remerge. These sentiments were expressed in their eyes, in fleeting glimpses. Our gloved hands could not offer the touch of comfort, but our eyes told them, 'You are not alone in this Covid ICU or ward. You might not have your family here, but you have us.'

Often, tears would shine in those eyes. But the gleam in them on being discharged, made our day. I can never forget the eyes of the staff nurses who worked tirelessly in those PPEs for long back-breaking shifts, lifting and turning patients

UBUNTU: I AM BECAUSE WE ARE

to maintain oxygen saturation, suctioning, administering medicines, taking orders from doctors and often being screamed at by the very patients they were helping. Not once did I see anything in those tired eyes but compassion, empathy and dedication.

We also encountered relentless optimism in the faces of family and caregivers of Covid patients, on digital screens, in the daily updates of loved ones they could not visit. Their voices were sometimes breaking and vulnerable, at other times angry and frustrated. We adapted to this new normal Facetime with them.

Selfishness & Greed: These two are among the most dangerous aspects of human behaviour, yet they are widely prevalent. On one side there are the many truly altruistic ones, who offer selfless service, philanthropists who empty their wallets to feed the poor and hungry, and on the other, those who display selfishness, greed and insensitive behaviour.

Many patients often lie to us about their positive Covid status when being admitted for treatment, delivery or a surgery. It is only later, when the tests done at our own centre come back Positive, that we discover the deception. Worse is that they do not feel the need to apologise or take responsibility for such behaviour. The protocols are thus necessarily tweaked to eliminate 'trust'. The actions of some, impact all. It reflects how self-centred some people can become in a crisis; to save one, they do not mind putting many healthworkers at risk. I often reprimanded them for this, but I do not know if my words had any effect on changing their mindsets.

Similarly, there were many who broke Covid norms by not wearing masks. When caught by the cops, they argued belligerently that they lived in a free country, and they even spat in their faces. Others were detained for rioting, urinated in corridors, and swearing at doctors. Some even took joy in pelting with stones and debris the doctors who came to test the population. Some hid their Covid status from their house-help so the latter would continue to come to work. Humans are capable of all this, and worse.

Hoarding groceries at the start of the lockdown could possibly be attributed to fear and panic buying, despite the Prime Minister's assurance that there was ample food in the country. But hoarding of medicines and lifesaving oxygen

UBUNTU: I AM BECAUSE WE ARE

cylinders merely reflects the depths of human behaviour.

No wonder then that nature itself has wreaked vengeance on humanity.

Love: The one emotion sans which we would all be robots, was seen blooming in many subtle ways. It helped us endure the lockdowns. Spending more time at home was a blessing in disguise and brought us closer to our children, parents and spouses. I am sure those children whose parents often return late from work, felt more secure and loved having them at home. Absence makes the heart grow fonder and social distancing made us realise more how much we missed hugging our friends or having crazy family reunions. These turbulent times also revealed who our true friends are – the ones who stood by us in sickness and in health, sharing our pain or just helping us stay cheerful. While others, who often posted Likes and Thumbs ups on social media, turned out to be superficial and unreliable when times were troubled.

There was also a surge of pandemic *pawmance*, with pet adoptions. Pets proved to be therapeutic for many coping with anxiety, loneliness and mental health issues. Cuddling pets also releases oxytocin, the happy hormone.

Laughter: This was the common prescription I tried to ink for all my friends, family, patients and colleagues during the pandemic. Because there was so much gloom all around, we often forgot to laugh. Finding humour in day-to-day things is a gift. It is like a windshield wiper that doesn't stop the rain but clears the glass to allow us to see and keep going. Whether watching comedies or sharing lighthearted moments with family and old friends, laughter was a balm to our wounded souls.

Peace: When you cannot go outside, go inside. People did this in different ways, through meditation, yoga, books, music, or simply watching the world go by for a change instead of chasing it. Moms yearned for some peace from their kids, waiting eagerly for schools to start, while couples signed peace treaties and kitchen pacts. *Do not let the behaviour of others destroy your inner peace*, said the Dalai Lama.

While the world fights the scourge of Covid 19, we need to fight our inner demons, moving away from greed, pride, selfishness and indifference.

We must collectively become a more sensitive, responsible and helpful society, in big ways and small. Covid *will* end one day. I hope we can also rid ourselves of pretentiousness and hypocrisy along with our face masks and return to our core humanity.

My favourite song by Michael Jackson is *We are the world*. It seems so apt in the present crisis.

There comes a time when we heed a certain call,
when the world must come together as one

There are people dying and its time to lend a hand to life

We can't go on pretending day by day

That someone somewhere soon make a change

We're all part of God's great big family,
and the truth you know love is all we need,

We are the world, we are the children, we are the ones who make a brighter day so let's start giving.

VAXENTINE DAY

We were standing on one side of a massive river of uncertainty and hardship…we are now seeing the other side of the river. ~ Christine Lagarde, President European Central Bank, on Covid vaccine.

14 February 2019, was *Valentine's Day.* A year later, the date marked *Quarantine Day,* as the Covid pandemic bought the world to its knees. But 2021 brought a shot of hope with the arrival of the much awaited Covid 19 vaccines. Perhaps 14 February 2021-22 will be known as *Vaxentine Day. Gift your loved one a shot of hope* could be the tagline.

As I brew my morning dose of caffeine, my better half filters the relevant news of our nation from the newspaper. 'So what's brewing in the country I ask?' 'Vaccine for nation,' he answers briefly. I then ask the usual daily question, 'What do you want for breakfast?' 'Slice of your life,' he quips. For the uninitiated, that is the title of my blog. Now there's a demure demand. When he has the loaf he asks for a slice? I serve him *sabudana wadas.* He complains they are 'spiky'

UBUNTU: I AM BECAUSE WE ARE

carbs sans the protein. I would have you know, dear reader, that I have long since developed a robust immunity to his remarks.

'So what about us?' he asks. 'Yes, our relationship is vaccing and whining,' I riposte. 'I meant what about us taking the vaccine?' he retorts, unamused. 'Are you done with your own interim analysis from your collected database or do you still need more time for peer review (those already vaccinated).'

Now, being a Paediatrician, I know more about vectors and immunology than an Ophthalmologist, who does not have an 'eye' for it. 'You go ahead and take the shot,' I tell him. 'No, I'll wait,' he declares. *In sickness and in health, and in vaccine,* might become the new normal wedding vows. I am dumbstruck by this proof of commitment from my better half. Has Covid perhaps affected him too?

As a Paediatrician, I am used to encountering fuss and reluctance from kids and parents in taking vaccines. Who likes to be poked with a needle? But Covid brought a new normal, with people not only pre-booking their shots but going a step further to make advance payment. This was in sharp contrast to the non-Covid era, when fees were left pending, and sometimes never paid. By now, even my house-help Shanta, knows names like Pfizer, Bharat, and Serum! She also knows Adar Poonawalla (she always smiles when she mentions him).

My Gujarati neighbour Pareshbhai, who is a SSC pass sharebroker, has developed a sudden interest in science. The other day he told me that Covid causes nervous system side effects like *Gira bar bar*. This made me nervous until I realised he meant Gullian Barre Syndrome. I doubt if most of us have ever studied so much about genome, viruses, RNA and DNA, even in college. India can certainly claim to have achieved 'heard' immunity, if not 'herd'.

Choosing A (Adar) over B (Bharat), is truly like selecting *Diwali* sweets. What to indulge in? You desperately want to eat but are scared to eat the wrong thing and then face consequences. Isn't it ironical that when the pandemic gripped us, we prayed for the vaccine, but now that it is at our doorstep, we shy away? Humans are a fickle race.

As doctors, my spouse and I were on the first list of vaccine recipients. While the rest of the nation got to decide from our experiences, we healthworkers led by example. Would I be labelled 'Designated Survivor' after the shot I wondered?

So there are two options. Covishield by Serum Institute, and Covaxin by Bharat Biotech. It is like Draupadi being asked to choose between Arjun (Covaxin) and Karna and his *kavach* (Covishield). Both are brave warriors, both good archers, both robust. But who would last longer and be the better husband? Draupadi can only tell after marrying one or the other. Similarly, which vaccine gives the longer immunity? *Khao toh jaano* (take it to know it). The elites want the imported Pfizer and Moderna shots. Despite the Prime Minister's *Atmanirbhar* campaign, their *atma* is always in foreign lands.

People ask me which shot to take all the time. The jury is still out.

Vaccine Venue Scenario: First Day First Show

Some enthusiastic and brave colleagues went on the very first day to get their shots. Others thought it best to get it done before the vials got exhausted (in true Indian fashion, the supply chain is always in question). Others got restless that they were not in the first lot of pokees. Relax.

Immunogenicity also seemed linked to taking a selfie while being injected, posting it on social media, along with video clips from the waiting area, and then blogging about the experience. I wanted to say it was barely a pinprick, that making 2-minute noodles was more taxing. But I held my peace. Everyone is different after all.

The waiting room is filling with mixed feelings. Some have cold feet, as if they are going sky diving, others breathless, some cannot control the excitement and adrenaline rush. Still others take prophylactic avil and paracetamol, or practice yogic breathing. The casual ones pose for pictures.

Then the dress code. Vaccine transparency is not important here. Ladies are in sleeveless or short-sleeved tops for convenience. The men happily flaunt vests and off-shoulder shirts. Those who did not have the forethought to wear short sleeves and loose fitting shirts, strip partially, flaunting hairy chests.

The thirty minutes post-jab status update is a must on WhatsApp: 'I am alive and kicking.' 'I am able to drive my car.' 'I am able to pee.' There are post-jab parties to celebrate the milestone event. My husband and I were however uninvited to these, as entry was only for the jabbed. Only non-alcoholic beverages and non-vegetarian jokes are served. I am still puzzling out the science behind refraining from alcohol for four weeks post vaccine. Shot in the arm but not in the glass?

All said and done, I feel proud of my country and the Government for efficiently managing the pandemic of 2020. India is also the proud manufacturer of two effective Covid vaccines. We even offered it to friendly and not-so-friendly countries. As citizens, we should not be hesitant about taking the vaccine, thinking Covid has reduced in the country. The second wave was devastating and the virus is mutating all the time. We need to protect ourselves. The Government has set a target to vaccinate the entire adult population by end 2021. So let us all join hands in this vaccination drive so we can meet and hug again, so children can return to school, so we can have normalcy in our lives again. Let's *vacci-our-nation* and yell, *Go Corona Go! Its time to roll up your sleeve for a shot of hope in the arm. So what are you waiting for?*

'So how about going on a vaccine date?' my better half asks. 'Co-vaxinate,' I reply, ending my own dilemma.

Have you? No ifs or butts, just a shot in the arm.

MAID OF HONOUR

*Here's to those who inspire
you and don't even know it.*
~ Anon

'Mom, who do you love more?' My kids often bowl this googly at me. My son claims that being firstborn, I love him more, while my daughter professes that the youngest is loved the mostest. In this love contest the husband gun fires and declares it is he I love most. By now I am a seasoned batsman at home in facing such spin attacks. This time, however, I declare out of the blue that I love my maid the most. She is my lifeline and without her my life goes on ventilator. At award functions I am sometimes asked to name the person behind my success. Without hesitation I say Shanta and Laxmi, my two maids. Behind every successful woman there is another woman - her house-help.

Lockdown, Scene 1: My alarm (the sound of clapping and cheering), goes off at 6am. I am forced to rise to the applause with a *yay, another day* feeling. I find my way to the kitchen with

UBUNTU: I AM BECAUSE WE ARE

eyes semi-closed, light the gas after a few failed attempts due to poor hand-eye coordination, and so kickstart my day. When you use your laryngeal muscles at the highest pitch to wake the kids, the hands to chop onions, the mind to decode diagnosis of patients admitted the previous night, the legs to sprint from one room to another, breakfast plates in hand, it is called 'multiorgan multitasking'. This is my morning routine in Covid times.

The Before Covid Scene: But my life was not always like this. Every morning, at 7am, the doorbell would ring and my maid, clad in a fresh sari, flowers in her hair, would stand flashing her million rupee smile. She would kickstart my day with a hot cup of *masala chai*. After gym, she would fix me a cup of my favourite filter coffee, with some hot breakfast. In lockdown, all that seems like some sepia-tinted memory. Once upon a time in Mumbai...

When lockdown happened, housing societies stopped domestic helpers from entering the premises. The only ones left were the live-in staff. *Atmanirbhar* became the new normal and emulating our Western counterparts, we Indians enthusiastically took charge of home chores.

Celebrities flaunted pictures on social media of them washing utensils or broom in hand. But the grass on other side proved not to be green at all, on either side.

With unemployment, life turned into a nightmare for domestic workers. While some employers continued to pay salaries to their domestic help during lockdown, the majority of such workers were left to fend for themselves. Unable to pay rents and electricity bills, many moved back to their villages, where life was safer, but not surer. For many, even meeting ends or providing the next day's meals for the family became daunting tasks. 'I would rather die of Covid at work than die of hunger at home,' I heard over and over again.

Post lockdown, many former employers were reluctant to hire them back, fearing the virus would travel from the overcrowded slums and hutments, bringing a high risk of Covid transmission. But nature has its own protection and it was the affluent in Mumbai's highrises rather than the poor in its slums who bore the brunt of the brutal second wave, due to better herd immunity among the latter groups.

Those who returned to work faced other forms of harassment, with societies imposing irrational terms and conditions. While domestic workers took care of children, cooked and cleaned, they were forbidden to use lifts and water coolers. Climbing ten or fifteen floors in highrises was brutal on them. In the days when the caste system prevailed, some were segregated from the rest. In Covid times, we have a new untouchability. We have evolved lifestyles but not evolved thinking. In Some buildings asked for Covid RT PCR reports every fortnight, which many employers refused to pay for. Some residents even went so far as to hide their own Covid Positive status from their maids so they would continue to come to work.

After a few months of lockdown, I had reached a peak of exhaustion, juggling between Covid duties and home chores. But when unlocking happened, my maid did not return from her village, despite my having continued to pay her salary through the difficult months. So the search for a new maid began. I wondered why newspapers did not carry ads for domestic work as they did for matrimonial. Something along the lines of: *Wanted a dedicated, hardworking, compassionate, ever-smiling, female domestic help age 25-35yrs, unmarried, no children, able to cook all cuisines, do dishes and laundry. Not interested in screen time.* Alas I had no luck. I broadcast my search on all possible platforms – WhatsApp groups, my children's PTA, family on both sides, medical association groups, Twitter, neighbours, watchmen etc.

The next few days were spent interviewing maids. This was one time the whole family showed solidarity. While interviewing, each of us had questions regarding our individual requirements. My son wanted to know if the candidate could brew fresh coffee like CCD. My daughter asked if she could iron and cook pasta. My husband enquired if she could do dishes without leaving grease. The questions became so individual specific that I wondered if we should perhaps look for four maids, one designated to each member of the family. Customized maids.

Finally, we found one who made our lives easy for a few months, till the second lockdown. Once we had made the pick, there began the MIP (maid internship programme), including a guided tour of the home and training. It was like training a Houseman to secure his first IV line (vein in vain). The first few days were like an observership – I did all the work and she got paid to observe.

The only positive outcome was that I lost a few more pounds. In fact, my husband insouciantly mentioned that he preferred no maids in the house since we were doing so well. Taking a deep breath, I reminded him that actresses playing double roles were paid double. Here, in real life, I was playing all roles for free.

According to official estimates, more than four million people are employed as domestic workers, often for less than subsistence wages. The unofficial number is closer to 50 million, of which two thirds are women. Shanta, who cleaned my clinic, has two children. She is the sole earning member of the family. Her husband is an alcoholic and often physically abuses her. She earned about INR 10,000 before Covid, and now hardly INR 2000 a month. While it remains a system of mutual dependency, people are afraid to have them in their homes. Ironically, people are not afraid to go to restaurants, rip off their masks and dine, or wander around malls with their masks dangling below their chins.

The truth is the virus does not differentiate. We might have the choice to stay at home and do without help, but they have no option. For them, it is survival. In fact, the first Covid case was reported in Mumbai's Dharavi slum, a domestic who worked in the home of a frequent international traveller, who had brought it in.

If we study the unfolding spread of the pandemic, it was where people went to events unmasked, did not follow protocol in social spaces, hosted weddings, or frequented religious and political gatherings, that the spread was rampant. Our house-help is not only masked but their voices silenced as well. They cannot vent on Twitter or TV. The treatment of domestics shows our society in poor light. We are neither liberal nor modern. Nor are we just.

Just as our house-help comes into our homes to clean and cook and care for us, it is high time we cleaned our hearts and minds of the lingering debris of discrimination and filled that space with compassion and kindness.

Take care of Lakshmi when she enters your home.

UPDATE YOUR STATUS

True wealth is not measured in money, status or power but in the legacy we leave behind for those we love, and those we inspire. ~ Cesar Chavez

I sit in my consulting room. Nowadays, practice is more at the spinal rather than cerebral level. A couple brings in their three-month-old infant. I am the fourth doctor they have seen, a typical case of doctor shopping. I enquire about the problem. They give me the list of differential diagnosis they have received thanks to Dr. Google, Moms WhatsAap Club and Cool Dads FB Group. 'Doc, our baby has Sandifer Syndrome. He has been vomiting for two months.' I sigh inwardly. Working in a Covid ICU and Covid wards for over a year, some of us have developed FOF (fear of forgetting paediatrics). I excuse myself and sneak into the next room and open Nelson's textbook of Paediatrics to read about Sandifer disease even though I know this to be a simple case of Gerd reflux.

Meanwhile, my cousin, who has been admitted for mild Covid, calls to ask why the doctors are not treating him with a monoclonal antibody cocktail for speedy recovery, like Trump had. Thanks to various WhatsApp universities, the whole world knows everything about Covid, from diagnosis to treatment.

Ironically, the majority do not know or do not care about wearing the right type of mask, in the correct manner, how to dispose of masks and shields, or even the right way to wash hands. My WhatsApp is filled with images of soiled diapers, patient's body parts, pictures of oral cavities, etc. In fact, I could easily compile an album. When the pandemic began, and lockdown happened, social media usage escalated by as much as 40-50%. It was a tool to communicate, to stay in touch in times of social distancing. For some, it served as a mood elevator. People pinned their interest to FB, Twitter, Instagram, and WhatsApp, while Zoom kept away the gloom.

One day, someone wished me happy birthday. Soon I had a flood of messages, that only stopped when I intervened to say it was not in fact my birthday! For many people emojis and thumbs up have become automode, without perhaps reading or knowing what they are thumbing up for.

The infodemic and tsunami of opinions on social media goes viral faster than the Covid virus. The overload of misinformation has been a nuisance, and also caused mental health problems. India has 350 million social media users, and a large proportion never check facts or trace sources of information. We are, it would appear, a gullible people. From quacks offering cures to herbal immunity pills, to religious prevention, racist and xenophobic posts, it has turned into a cesspool of fake news, civil unrest and political division.

Even before we had significant Covid cases in the first wave, fake messages on social media led to the hoarding of masks and sanitisers. As a result there was shortage of N95 masks for doctors. The same happened in the second wave with drugs like Remdesivir, oxygen cylinders, and availability of beds. Then came the anti-vaccine lobby, who even suggested menstruating women could die if they took the shot. Herbal alternatives to treat Covid were expounded. Social media as a place to seek health information is nothing but a fool's paradise.

Corona deaths, post-vaccine complications and vaccine myths have escalated mental health issues by 40% in the last year, with anxiety, panic attacks, depression, insomnia and more, taking hold of the susceptible. Even children display emotional and behavioural issues, or speech delay because of poor interaction at home, and virtual autism.

Facebook is La-La-Land. People portray what they think others want to hear in order to get a confetti shower of Likes and comments on their photos and updates. It often leads to comparison and complexes for many, with false illusions that others have better lives, are more successful, better liked. Why does no one ever post an aadhaar card photo I wonder?

Social media is both an ally and a potential threat. A major drawback is that it quickly disseminates false information, which can confuse and distract. There is an urgent need to take effective measures regarding use and misuse. Messages from authentic medical sources should be the only ones allowed to educate and create awareness in medical situations. It is time to create a multi-disciplinary approach and real time information sharing system with a team of experts, to draw data, analyse, and detect misinformation before it snowballs. The Government needs to accept that social media is part of our lives and must to be regulated like other areas.

And it is important to like oneself, while not necessarily being liked by others. Just as it is necessary to like what you get in life and not just crave what you like. And every morning, when you rise, make your self-appreciation and self-confidence as strong as your coffee.

CATCH ME IF YOU CAN

When I is replaced with Us, and Me by We,
illness becomes wellness. ~ Author

15th Aug 2021, we celebrated our country's 75th Independence Day. While I was praying in my heart for a world free of Covid, I did not know that the virus had already invaded my own body and that I was in the incubation period of a Covid 19 attack.

Seeing me sniffle, a colleague asked whether I was feeling unwell? Just allergic rhinitis, I said. By nightfall, I had fever and chills. I reassured my family (and myself), that it was nothing more than the viral flu so many of my patients were suffering from. After three days the fever rose, and the body ache was bone deep. The fatigue and weakness was so great I could barely walk.

'Let's just do a RAT (rapid antigen test),' said my microbiologist, 'just to be sure.' Five minutes later, the result was Positive. It had finally attacked me, the virus I had tried to keep at bay for 18 months, despite spending hours in the Covid ICU and Covid wards, and treating hundreds of Covid patients. When disease enters your body, it feels like the enemy invading your home.

Over the next 24 hours, my symptoms worsened. I became dehydrated and my BP dropped. I was admitted to Reliance Hospital, Navi Mumbai, where I am a Consultant, in the same ward I had briskly and confidently walked each day and asked in my cheerful tones, 'So how are you today?' The tables had turned overnight. I was now the patient, frail, febrile, and sapped of energy lying in bed. From wearing a doctor's blue scrubs to wearing the patient garb of blue shirt and pyjama, I thought how unpredictable life is.

The team of doctors and nurses were all on their toes – iv line secured, labs sent, fluids pushed, vitals checked, ecg leads, monitors attached and medications pushed

Did I feel powerless? No I did not. I felt humble and touched by the kindness and care shown by my colleagues. I lost all sense of ego and felt like a drop in the ocean of this world. More importantly, I felt at home thanks to my hospital team, who cared for me like a family member. Small things felt so wonderful, like the duty nurses being careful not to disturb me if I was sleeping, to take vitals, or housekeeping bringing food when I wanted, the chef asking if I had any special request, duty doctors reassuring me, and friends and colleagues coming to meet in PPEs, telling me I was not alone. *Ubuntu* – I am because we are. I understood its true and deep meaning in those moments.

Did I feel fear? No. For I had seen it all in the last year- patients crashing in ICU, going onto ventilators, bleeding crisis, sepsis and death. But I did not feel afraid. I trusted my doctors more then I trusted myself. And my mind was sturdy like a fortress which no negativities could invade. I had to recover for I am not done with my life. I still have miles to go before I sleep.

The rest of the family also developed mild symptoms and were quarantined at home. Since both children were running fevers, I took early discharge from the hospital and returned home. I forced them to stay hydrated. By then, we had all lost our sense of taste and smell. With no domestic help at home, it was a difficult time. But tiffins were dropped at our door daily by family and close friends, as were medicines. Wishes and prayers poured in.

Covid sucks you of all energy. Even a week later, when the fever had subsided, we still felt exhausted. We had no appetite. We ate only because we had to. I yearned to get back my sense of taste and realised the true value of the normal things of life that we take for granted. Often, we just ate in a hurry, watching TV; mindless eating without savouring taste or texture. You have to lose it to realise the value.

Can't say if being Cupid or Covid struck is worse. Both make you go weak in the knees. While one deprives you of taste and smell, the other deprives you of all your senses. Fortunately, I did not lose my sense of humour.

We slept for 14 to 16 hours a day. Our home was like a doctors on-call room – sleep, eat, sleep again. We lost sense of time and date. The human body is such a marvel in its natural ability to regenerate and heal. Sometimes, falling ill is our body whispering to us to slow down, that we need rest to heal, for renewal of cells, and regeneration of energy of body and mind. Retrospectively, I realise the last 18 months as a frontline warrior, I was working hard to keep pace with my Covid and non-Covid patients, without a break. Maybe this enforced rest was what my body needed. A vacation in the mountains would have been better, of course. But it is a time to accept whatever comes, to survive and remerge braver and stronger.

So we decided to consider it bonus family time, when having three meals together every day was indeed a blessing. Other than Sundays, we seldom had that opportunity normally. We looked after each other, watched old movies, helped each other with household chores, and shared our Covid woes and laughed.

Self-pity is the worst thing in such situations. So even on the days we did not feel so good, had nausea, could not eat much, we still buoyed each other's spirits. It felt stronger together. Friends kept us cheerful by phoning every day.

Weeks passed. Slowly and steadily we moved towards normalcy. Taste and smell took more than three weeks to return. Covid fatigue remained and will take longer to go. But hey, no complaints. The experience has made me value my life, health, family, friends and profession all the more. I would always laugh and fiercely declare that Covid 19 could not catch me.

How wrong I was.

But I have no regrets, only gratitude for our recovery. For all the good people in our life, who stood by us. Together, the human spirit triumphed yet again.

EPILOGUE

REBOOTING LIFE

*You are saying the caterpillar is being pushed
I am saying the butterfly is born.*
~ Mahatria Ra

'Mom will you consider this a bad year?' asked my daughter. 'No, honey. It is always good or not so good, but never bad,' I answer. The pandemic upended our lives, changed our routines, life goals and priorities. But it also made us agile and adaptable to a new normal. And if we miss some of the comforts we had before, it was nonetheless a good year and many of us felt privileged to have had more family time. We learnt to be more independent and also realised the value of mutual dependency, resilience and humility.

We learnt that the health of our planet and the health of human beings are interlinked. The Pangolin virus, transmitted via bats, caused the whole pandemic, highlighting the illegal trade of various species. The illegal trade in animals is the fourth largest trade in the world. Research revealed that countries with low pollution levels had milder cases. And we all witnessed clear skies and clean air because vehicles were off the roads. There were green lawns, monkeys strolling the roads, less crime, and less infectious diseases. All benefits and lessons of the pandemic. We slowed our pace of life and no longer ran on life's treadmill. Finally, we had time to stand and stare.

We learnt new skills – cooking, sewing, painting and music. Personal benefits for me were reading and writing. We realised we did not need gyms to stay fit, own homes were good enough.

Our mutual dependency, be it with humans or the ecosystem, was a lesson well learnt. It improved our EQ, and fostered kindness and gestures of solidarity. There was time to contemplate and prioritise things in life.

Perhaps the most important lesson was living with an attitude of gratitude. The word 'gratitude' comes from Latin *gratus* (pleasing, thankful, feeling of appreciation felt or shown, gratefulness or graciousness), signifying a positive mindset and actions. It's not happiness that makes us grateful but being grateful.

Is it really true that happy people are grateful? We all know some people who seem to have everything yet they want more, or something else. On the other hand, there are those who face misfortunes but still have a sense of peace and contentment. Finding gratitude during the trials of life, in death and disease, is difficult. But embracing those trials with gratitude gets you through them.

OTHER LESSONS TO TAKE TO HEART:

Count your blessings: Stop focussing on what you do not have.

Embrace humility: A humble heart is always more contented with simple things.

Be minimalistic: The pandemic surely taught us that less is more.

The current economic recession will improve and we should allow for no recession of the human spirit, or our zeal for life. You can still play with your children even if you cannot go on vacation, lie down and look at the night sky, train employees, foster leadership, reboot health, knowledge, and spiritual growth. The pandemic will end and when it does, do not look back and see

UBUNTU: I AM BECAUSE WE ARE

that you did nothing but wallow in gloom. When toxins accumulate in the body, we die. Similarly, if toxins accumulate in our minds, they kill us just as surely. If you hold a glass for too long, the hand starts to hurt, so we put down the glass. In the same way, let go of grudges and anger.

True freedom is not being concerned about other people's opinions, views and actions. You do not need to explain or justify. Free from ego, one is truly free. When we cease to be 'I' specialists, we can 'see' other people with greater compassion and less judgement.

It is only in the crucible of life's challenges that Man is pushed forward and humans evolve. And when this pandemic ends, we will once more break free from our cocoons and enjoy the flight of this beautiful life, but with a new respect for the world we live in, and the priceless gift of our own freedom.

ACKNOWLEDGEMENTS

*We must find time to stop and thank the people
who make a difference in our lives.*
~ John F Kennedy.

This book originated from conversations I had, and blogs I penned, during the lockdown. The ideas were nursed in Covid ICU and Covid wards, hospital corridors, in shopping for essentials, and at home.

I would like to thank the entire team at Leadstart for once again shaping my manuscript into a book. Swarup Nanda, CEO, and Malini Nair, Head of Publishing, for believing in me, and for all their support. Chandralekha Maitra, my Editor, whose eye for detail brings magic to the book with her brilliant editing. Ananya Subramaniam, my Relationship Manager, for all the constant updates.

My better half, Prashant, who doesn't read my books but jokingly justifies this by saying he reads me. I playfully ask him whether he reads me 'between the lines' or 'thoroughly'? It has

UBUNTU: I AM BECAUSE WE ARE

been a long journey since college days, of mutual support and encouragement, helping each other grow to be our best versions.

My two grown children, Saahil and Kavya, who invariably barge in when I am writing at home. When I point this out, they riposte that I can think and plot a story in my mind anywhere, even in the bathroom or kitchen, that I do not need the quietude of a writers den. I take that as a positive endorsement.

My parents, for my education, not only in medicine but also in preparation to journey along life's road.

To my biggest teacher of 20-2 – the Sars Covid 19 virus. An eye-opener indeed. It made us revisit our plans and priorities, to reflect and contemplate. Most importantly, it made us realise that health is wealth.

My little patients, who have always taught me life lessons beyond books. I am perennially thankful to them, and to their parents, for trusting me with their children's lives.

My friends from medical college, who are both my honest critics and staunch supporters. They always add cheer and fun to my life. My 'sol' buddy, who cherishes my hopes, helps fulfil my dreams, supports my endeavours including some stupid and impulsive ones, and partners me in joy, sorrow and mischief.

To all the wonderful authors and the books I have voraciously read, which have nourished my mind and soul.

To the Almighty, for this beautiful life. At the end of my days, when I stand before Him, I hope I will be able to say in truth: *I used everything you gave me.*

ALSO FROM SHILPA AROSKAR

Using riveting real life stories about the triumph of the human spirit over adversity, **YOLO** presents a relatively unexplored perspective of medical treatment and recovery. It highlights the mind-body and link, mystical healing, spontaneous remissions, reinforcing the greater truth that the human mind is perhaps the biggest pharmacy and the power of faith and hope can indeed move mountains.

Drawing from her rich professional experience and personal life, Dr. Aroskar addresses fundamental questions like: *Why are some people more prone to disease than others? Why does the same treatment heal one and fail another? How can one remain relentlessly optimistic and resilient when confronted by daunting challenges? How does one stay strong in the face of adversity? The doctor-patient relationship,* and *the dilemma of euthanasia.* Despite epochal advances, medicine remains an imperfect science.

A deeply engrossing and enlightening book.

ABOUT THE AUTHOR

DR. SHILPA AROSKAR is a paediatrician by profession and a writer by passion. She is currently Head, Department of Paediatrics and Neonatology, Reliance Hospital, Navi Mumbai. Gifted with the human touch, she has always gone beyond being a clinical professional to devoting her efforts towards being a force for good. Whether as President of the Indian Association of Paediatrics, Navi Mumbai, or working with less privileged communities, her focus has been to create sustainable impact.

An avid reader, she first took to writing as a stress buster. It soon became a passion. Her book, *Parents Guide to Child Care* is considered the Bible of Indian parenting. Her second book, *YOLO: You Only Live Once* drew inspiration from her rich professional-life experiences. Her popular blog: Slice of my daily life@drshilparoskar on Word Press, draws readers across age groups, ethnicities and geographies. In her trademark style, she makes her readers laugh, ponder and wonder at the commonalites and mysteries of life.

Married to an ophthalmologist, and mother of two, Shilpa lives in Navi Mumbai. She loves to travel, seeing the world as an exciting adventure

ABOUT THE AUTHOR

book to be savoured. Her other interests include painting and music.

Contact: contactssparshclinic@gmail.com

Website: www.drshilparoskar.com

UBUNTU: I AM BECAUSE WE ARE

9 789354 588303